PROGRAMMED INSTRUCTION HANDBOOK

NONDESTRUCTIVE TESTING

Introduction

Published by *PH D*iversified, Inc.
5040-B Highway 49 South
Harrisburg, NC 28075

Copyright © 1995 by
*PH D*iversified, Inc.
Second Printing February 2000

All Rights Reserved

No part of this book may be reproduced in any form
without written permission from the publisher.

Printed in the United States of America
ISBN 1-886630-00-3

ACKNOWLEDGMENTS

Publishing and Printing

    Revision Editor: . . . . Dr. George Pherigo, **PH D**iversified, Inc.

    Production Editor . . Ms. Mary Lou Hollifield, **PH D**iversified, Inc.

    Proofreading . . . . Ms. Jean Pherigo, **PH D**iversified, Inc.
    Proofreading . . . . Ms. Dana Smilie

Technical Content Revision

    Technical Editor . . . . Mr. Robert W. Smilie

This handbook was originally prepared by the Convair Division of General Dynamics Corporation under contract to NASA and was identified as N68-28776. This book is the first in a series of books, commonly know as the General Dynamics Series, that has been the basis of many industrial NDT training programs for over 20 years.

Now, after several decades of widespread use, the entire series has undergone a major revision. The revised material no longer concentrates on applications in the aerospace industry, but instead, covers a wider range of industrial applications and discusses the newest techniques and applications.

Mr. Robert W. Smilie has been the principal author of the revised material in this text. Using his nondestructive testing experiences in several industries, including work at the EPRI NDE Center, he has updated the text to better suit the entry-level technician/engineer.

TABLE OF CONTENTS

                                                         Page

Preface iv
Instructions v

Chapter 1 - The Need for Higher Quality 1-1

    Quality Through Nondestructive Testing 1-13
    Destructive and Nondestructive Tests 1-18
    Discontinuities 1-19
    The Origins of Discontinuities 1-23
    Chapter Review 1-56

Chapter 2 - Working the Billet 2-1

    Laminations 2-6
    Stringers 2-15
    Seams 2-17
    Chapter Review 2-38

Chapter 3 - Forging Discontinuities 3-1

    Forging Laps 3-4
    Forging Bursts or Cracks 3-22
    Chapter Review 3-33

Chapter 4 - Casting Discontinuities 4-1

    Cold Shut 4-5
    Hot Tear 4-6

Shrinkage Cavity	4-24
Microshrinkage	4-31
Blow Holes	4-39
Porosity	4-44
Chapter Review	4-47

Chapter 5 - Tubing, Pipe, and Extrusion Discontinuities 5-1

Welded Tubing and Pipe	5-1
Seamless Pipes and Tubes	5-11
Extrusion Discontinuities	5-18
Chapter Review	5-23

Chapter 6 - Processing Discontinuities 6-1

Grinding Cracks	6-1
Heat Treating	6-4
Explosive Forming	6-15
Fatigue Cracks	6-18
Chapter Review	6-29

Chapter 7 - Welding Discontinuities 7-1

Welding Terms	7-2
Crater Cracks	7-3
Stress Cracks	7-16
Porosity	7-19
Slag Inclusions	7-28
Tungsten Inclusions	7-33
Lack of Penetration	7-37
Undercut	7-45
Lack of Fusion	7-46
Chapter Review	7-52

Chapter 8 - Other Engineering Materials 8-1

    Structure of Materials 8-2
    Plastics 8-13
    Composites 8-16
    Ceramics 8-26
    Powder Metallurgy 8-30
    Chapter Review 8-35

Introduction to NDT Self-Test A-1

Glossary B-1

PREFACE

Programmed Instruction Handbook - Introduction PI-1, is one of a series of training handbooks designed for **self-study** applications. The programmed instruction format allows the student to learn the material when a formal classroom setting is not available.

It is recommended that this **Programmed Instruction Handbook**, PI-1 Introduction to Nondestructive Testing, be completed before starting other books in the series. The basic information about metallurgy and discontinuity types contained in this book will make other books in the series more understandable and easier to master.

This **self-study format** provides self-evaluation quiz questions at the end of each chapter and at the end of the book. A score of 80% or better on the self-test will indicate you understand the basic Level I concepts on this subject.

This **Programmed Instruction Handbook** is also very helpful when used prior to or in conjunction with any of the **Classroom Training Handbooks**.

Other **Programmed Instruction Handbooks** in the series include:

PI-2		Liquid Penetrant Testing
PI-3		Magnetic Particle Testing
PI-4		Ultrasonic Testing
PI-5		Eddy Current Testing
PI-6		Radiographic Testing

INSTRUCTIONS

The pages in this book should **not** be read consecutively as in a conventional book. You will be guided through the book as you read. For example, after reading page 3-12, you may find an instruction similar to one of the following at the bottom of the page:

- Turn to the next page.

- Turn ahead to page 3-15.

- Turn back to page 3-8.

On many of the pages you will be faced with a choice. For instance, you may find a statement or question at the bottom of the page together with two or more possible answers. Each answer will indicate a page number. You should choose the answer you think is most correct and turn to the indicated page. If you happen to select an incorrect answer, continue to read, as the page will provide supplemental information to help you understand the concept.

We know that sometimes the information in this self-study format may seem oversimplified or repetitious. Bear with us; the reinforcement of basic Level I concepts is essential if you expect to retain the knowledge and apply it to Level II training or on-the-job NDT applications.

Do not rush through the volumes. Take whatever time you need to make sure you have a clear understanding of the material. Depending on your background knowledge, reading speed, etc., the time needed to complete this book may vary from **5 to 10 hours** or more. As you will soon see, this self-study handbook is easy to use - just follow instructions.

TURN TO THE NEXT PAGE.

CHAPTER 1

THE NEED FOR HIGHER QUALITY

Consumers of today place an emphasis on quality because world markets demand it. International standards for quality have been in place for some time and to compete globally manufacturers must meet these minimum standards. In fact, purchasing decisions are often made on the basis of a product's performance history.

The products being marketed today must meet greater demands on reliability. This includes such things as cleanliness of the assembly area, proper manufacturing techniques, and the precise alignment of doors and fenders on an automobile, for example.

Product quality is necessary because of the value and performance expectations of consumers. We depend more and more on the reliable and consistent operation of everyday "consumer products." Our dependence on consumer products continues to grow, whether from the home or office computer, the facsimile machine, the cellular telephone, or the automobile we utilize to carry us to and from work or play. Because we demand long-term, trouble-free service, each subassembly that becomes a part of the final product must be designed and manufactured with quality in mind.

From the above, you can conclude that the customer is demanding which of the following? (Choose one and turn to the page indicated.)

A less expensive product	**Page 1-2**
Customer satisfaction	**Page 1-5**
A reliable product	**Page 1-8**

From page 1-1   1-2

Yes, we all demand less *expensive* products, but we do not want an *inexpensive* product that breaks down all the time. This represents a poor value.

The consumer that purchases a product demands that the product will give trouble-free service for a reasonable time period. If you purchase a cheap facsimile machine, you may find that it cannot be relied upon; it may break down constantly. However, an expensive product does not assure product quality.

Not all low-priced items are unreliable, and not all high-priced items are reliable and of good value or quality. The relationship between price and value (or quality) depends on the manufacturer and the price the consumer is willing to pay for a specific product.

Turn back to page 1-1 and try one of the other answers.

From page 1-8

When you consider the hundreds of thousands of critical parts used in the construction of a jetliner, the task of obtaining the desired level of reliability is monumental. What can happen if a fuel line breaks in a jet engine pod? What can happen if a flight control cable breaks or comes loose in flight? What can happen if the landing gear fails to extend for landing? The lives of all on board are jeopardized, and millions of dollars may be lost.

Considering the above, what do you think the customer is demanding now?

Customer satisfaction .... Page 1-4
100% reliability .... Page 1-6
A less expensive product .... Page 1-10

From page 1-3

You feel the customer is demanding satisfaction. Well, not much satisfaction can be obtained after a jetliner crashes and all of its occupants are killed. There is nothing to be gained by demanding satisfaction in this case.

No, the airline is demanding something more than satisfaction. It is buying a jetliner to carry several hundred passengers. Its first demand is that the jetliner function without fail.

Turn back to page 1-3 and study the problem again.

From page 1-1

All right. Let's say the customer is demanding satisfaction. But let's take a deeper look into just what that means.

Customer satisfaction stems from a product that is reasonably priced and has proven that it has a reasonable life expectancy.

For example, if the print quality of a color laser printer is not vivid and sharp, the consumer will be dissatisfied because the printer output will be less than expected. On the other hand, if the printer no longer operates after 500 or so copies, the cost per copy becomes high, along with the ill feelings of the consumer.

Most customers do not want to buy a product that has to be continually repaired - even if the repairs are free. So you see, the customer is demanding something more than satisfaction.

Turn back to page 1-1 and find the correct answer.

From page 1-3   1-6

Good for you! That's absolutely right. The customer is demanding 100% reliability, and rightfully so, for the customer must protect his investment and his good name.

The highest product quality is needed to assure reliability, not only for jetliners, but also for space-launch vehicles and their associated spacecraft. Consider the thoughts of an astronaut in preparing for a space shuttle flight.

Turn to the next page.

From page 1-6

The astronaut knows that the retro-rockets must fire to bring the space shuttle out of orbit. The control system must function. If the space shuttle cannot be positioned for re-entry, it will burn up in the atmosphere during its return to earth.

If you were the astronaut preparing for orbital flight, what would be your primary concern?

I would hope everything works .... Page 1-9
I would hope for 100% reliability .... Page 1-11
I don't think 100% reliability is possible .... Page 1-12

From page 1-1

That's right. The customer is demanding a reliable product. For example, a family of modest means could not afford to buy a telephone that breaks down all the time.

In manufacturing, the problem of reliability is magnified many times. A manufacturer of modern jetliners is dealing with a product that costs millions of dollars and which involves the lives of several hundred people every time the jetliner flies. You can see why an airline company purchasing an airplane would demand a reliable product.

Turn back to page 1-3.

You bet your life you would hope everything works. A space shuttle is not an attractive coffin. If the retro-rockets failed to fire, that's exactly what you would have - a space coffin.

So, now you understand why we need greater quality in these days of rapid technological advancement. Consider the first telephone conversation weighed against intercontinental television and phone calls relayed by communication satellites. Consider the challenge of landing men on the moon, then Mars, as opposed to earth orbital flights. The only way these endeavors succeed is by producing quality space vehicles that are *reliable*.

If you were going to the moon in a spacecraft, you would certainly hope for 100% reliability.

Turn ahead to page 1-11.

You feel the airline is demanding a less expensive product. The airline certainly wants to get as much for its money as possible, but not to the extent of sacrificing the reliability of the airplane.

With millions of dollars invested and the lives of hundreds of passengers involved every time the jetliner flies, don't you think the airline would want the airplane to operate properly every time it flies?

Turn back to page 1-3 and try again.

From page 1-7

That's right. You would hope for 100% reliability.

When the thousands of individual articles required in the construction of a launch vehicle and its associated spacecraft are considered, 100% reliability seems like an almost impossible task. But even one tiny bit of reliability should not be sacrificed because of lack of conscientious effort on someone's part.

Right here is where QUALITY ASSURANCE enters the picture. YOU as a member of a QUALITY ASSURANCE TEAM have the opportunity to help make history. How effectively YOU do your job may well determine whether or not the product is reliable.

The National Aeronautics and Space Administration (NASA) has embarked on a program that is the greatest technical effort ever undertaken. Intensive scientific investigations are carried out in every field of modern technology. NASA employs weather and communications satellites, deep space and lunar probes, and orbiting observatories. The NASA manned space flights are recorded in history. The most ambitious flights remain for the future. As the distance of these flights increases, the reliability and quality requirements will increase fourfold. Thousands of problems must be solved, new technologies mastered, space environments charted, and unknown environments studied. These achievements will not depend on a gifted few; rather, they will represent, now and in the future, the sum of the contributions of each and every one of us.

Turn ahead to page 1-13.

From page 1-7

You don't think 100% reliability is possible. If you interpret this to mean "perfection," then you are right because true perfection is unattainable. But let's look at that statement from a new point of view.

Let's call our approach a "Zero Defect Program." What this approach actually means is that, once the desired quality level has been established for a specific article, we have ways of eliminating those articles that do not meet these standards. In this way, we can be assured that the article will perform as advertised.

Admittedly, there are many variables involved here, and a defective article will slip through once in a while. If the article is critical, the results may be tragic. Any way that you look at it, the least that can be expected of a critical article that is defective is that failure can cost millions of dollars in the loss of a space-launch vehicle or spacecraft.

In direct support of the "Zero Defect Program," we provide back-up systems or other ways of performing critical functions. For example, in a spacecraft, two or more methods of controlling attitude are provided, since attitude is of grave importance when re-entering the earth's atmosphere. These back-up or alternate systems are provided to increase the reliability of a spacecraft to as close to 100% as possible. If you were an astronaut preparing for a space flight, wouldn't you hope for 100% reliability?

With this in mind, turn back to page 1-11.

Quality Through Nondestructive Testing

We have discussed the need for higher quality in terms of the high cost of today's complex equipment and in terms of saving lives. We also pointed out the extreme importance of YOU and every member of the QUALITY ASSURANCE TEAM. Now let's discuss ways of attaining product quality through the use of NONDESTRUCTIVE TESTS.

Nondestructive tests are exactly what the name implies - methods of testing articles (parts or materials) for cracks or flaws without damaging the article. A doctor uses many nondestructive tests in a physical examination. He taps the knee to determine the condition of the reflex functions of the nervous system. When the doctor takes an X-ray to test the condition of the lungs, he is examining the inside of the body without damaging any part of the body.

Turn to the next page.

X-rays are widely used to examine the inside of articles. This nondestructive test is called Radiographic Testing. Here is an X-ray of an electrical connector used on a space-launch vehicle.

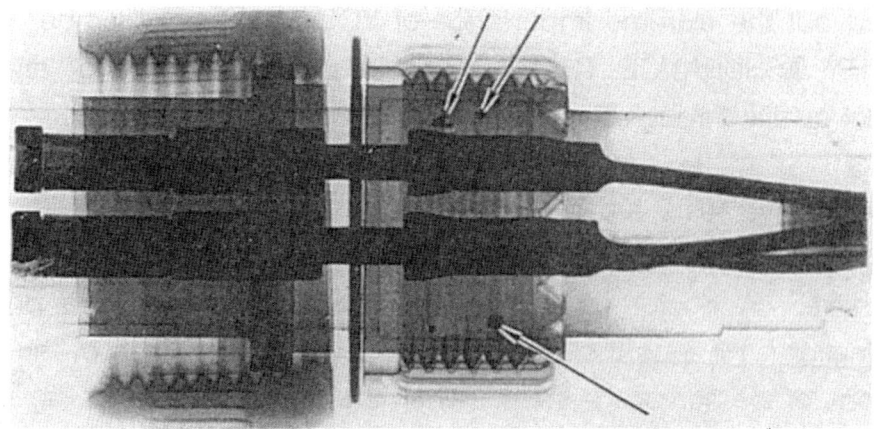

This article connects the circuit from the autopilot to the vernier adjusting boosters, which control the flight path of the launch vehicle. The arrows point to foreign objects found in the connector. These objects are pieces of solder that could short the connection, causing the launch vehicle to crash or fail to achieve the correct space flight path. Through the use of radiographic testing of the electrical connector, a catastrophe was averted.

Turn to the next page.

There are many methods of nondestructive testing to locate holes, cracks or breaks in the continuity of materials or articles. The nondestructive tests we will be concerned with are:

- EDDY CURRENT TESTING (ET)
- MAGNETIC PARTICLE TESTING (MT)
- LIQUID PENETRANT TESTING (PT)
- RADIOGRAPHIC TESTING (RT)
- ULTRASONIC TESTING (UT)

Notice the symbols in parenthesis at the end of each test method. These symbols are used on engineering drawings to designate the test method or methods to be used in determining the quality of a specific article.

Each nondestructive test is designed to provide visual evidence of flaws in articles not normally visible to the unaided eye. The visual evidence left by each method is called an INDICATION. The indication may be an accumulation of magnetic particles, as in magnetic particle testing. We see the accumulation of magnetic particles and not the actual crack.

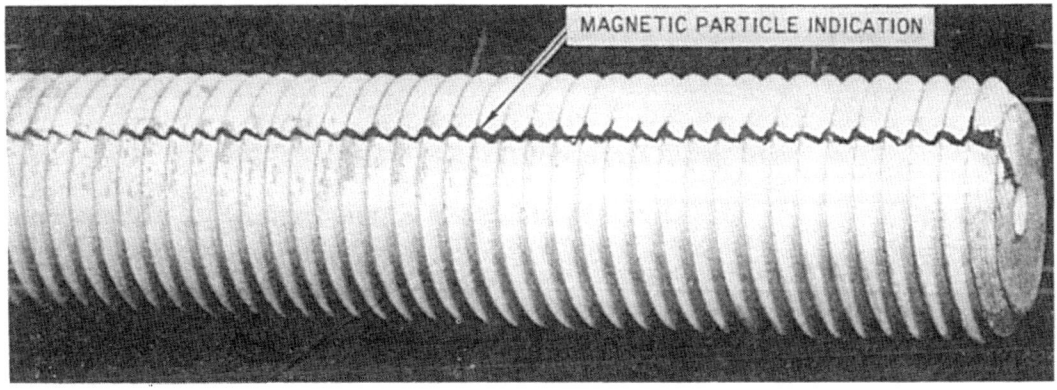

Turn to the next page.

From page 1-15

Other indications may be provided by an indication on a computer monitor for ultrasonic testing, or in the form of shadows or light spots on X-ray film in the case of radiographic testing. NONDESTRUCTIVE TESTS, then, are methods used to determine the performance capabilities of materials or articles without damaging them. If the test method is properly selected for a specific article and the test itself is properly performed, you can be sure that the article has no flaws and will meet the quality requirements.

Turn to the next page for a table that briefly describes the nondestructive testing methods.

From page 1-16

Table 1-1
NONDESTRUCTIVE TESTING METHODS

	LIQUID PENETRANT TESTING	MAGNETIC PARTICLE TESTING	RADIOGRAPHIC TESTING	EDDY CURRENT TESTING	ULTRASONIC TESTING
DEFINITION	USES A PENETRATING LIQUID TO SEEP INTO A SURFACE DISCONTINUITY THUS PROVIDING A VISIBLE INDICATION.	USES ELECTRICAL CURRENT TO CREATE A MAGNETIC FIELD IN A SPECIMEN WHILE MAGNETIC PARTICLES INDICATE WHERE THE FIELD IS BROKEN BY A DISCONTINUITY.	USES ELECTROMAGNETIC RAYS (X RAYS AND GAMMA RAYS) TO PENETRATE MATERIAL, RECORDING ON FILM DISCONTINUITIES IN THE MATERIAL	USES AN ELECTRICAL CURRENT IN A COIL TO INDUCE EDDY CURRENTS INTO A SPECIMEN. INDICATIONS REVEAL DISCONTINUITIES THAT ALTER THE PATH OF THE INDUCED CURRENTS.	USES ULTRASOUND TO PENETRATE MATERIAL, INDICATING DISCONTINUITIES ON AN OSCILLOSCOPE SCREEN.
USES	USED ON METAL, GLASS, CERAMICS TO LOCATE SURFACE DISCONTINUITIES. SIMPLE TO USE AND DOES NOT REQUIRE ELABORATE EQUIPMENT.	USED ON METAL WHICH CAN BE MAGNETIZED (FERROMAGNETIC) TO DETECT SURFACE OR SUBSURFACE DISCONTINUITIES. SIMPLE TO USE AND EQUIPMENT IS PORTABLE FOR FIELD TESTING.	USED ON ANY METAL STOCK OR ARTICLES, AS WELL AS A VARIETY OF OTHER MATERIALS TO DETECT (AND RECORD ON FILM) SURFACE OR SUBSURFACE DISCONTINUITIES. FILM PROVIDES A PERMANENT RECORD OF THE DISCONTINUITIES.	USED ON METALS TO DETECT SURFACE AND SUBSURFACE DISCONTINUITIES, HARDNESS AND THICKNESS, PLATING COATING (NONMETALLIC), AND SHEET THICKNESS MEASUREMENTS.	USED ON METAL, CERAMICS, PLASTICS, ETC., TO DETECT SURFACE AND SUBSURFACE DISCONTINUITIES. WHEN AUTOMATED, INDICATIONS ARE RECORDED ON PAPER, PROVIDING A PERMANENT RECORD. ALSO MEASURES MATERIAL THICKNESS.
LIMITATIONS	DOES NOT DETECT DISCONTINUITIES BENEATH THE SURFACE OF A SPECIMEN.	CANNOT BE USED ON METAL WHICH CANNOT BE MAGNETIZED. REQUIRES ELECTRICAL POWER.	HIGH INITIAL COST. REQUIRES ELECTRICAL POWER SOURCE. POTENTIAL SAFETY HAZARD TO PERSONNEL.	INSPECTION DEPTH LIMITED TO LESS THAN ONE INCH. DOES NOT GIVE PHYSICAL SHAPE OF DISCONTINUITIES.	MODERATELY-HIGH INITIAL COST. REQUIRES ELECTRICAL POWER SOURCE. INTERPRETATION OF TEST RESULTS REQUIRES HIGHLY-TRAINED PERSONNEL.

Turn to the next page.

Destructive and Nondestructive Tests

Nondestructive tests supplement destructive tests. DESTRUCTIVE TESTS destroy the article being tested. These tests usually bend, twist, or break the article being tested and destroy its usefulness for service. Since these tests destroy the article, this type of testing is usually costly and is limited to testing a small percentage of a group of specific articles. Destructive tests assume those articles not tested have equal quality. Such an assumption is not adequate for critical components, as each and every one of these must meet rigid standards of quality and reliability.

For this reason, nondestructive tests are used to supplement destructive tests. Nondestructive tests determine the qualities of a part without altering or changing its physical qualities or usefulness. Thus, each and every article is inspected without damaging its physical structure. Therefore, nondestructive tests supplement destructive tests and give further assurance that *all* articles meet quality standards.

Today's industrial designs are being pushed to the limits of available materials, engineering knowledge, and processes to produce required system components. More than ever before, reliability and freedom from maintenance are required of the end item. Nondestructive tests are an absolute necessity to provide all possible assurance that an article meets the required quality level and will perform as expected. Space, weight, or cost limitations often prevent over-design with the large built-in safety factors formerly permissible. Thus, nondestructive testing serves as a very important tool in QUALITY ASSURANCE.

Turn to the next page.

Discontinuities

Up to this point, words such as breaks, cracks, holes, and flaws have been used to show the type of things we will be looking for with nondestructive tests. All of these can be summed up in one word: DISCONTINUITIES.

The word dis-con-ti-nu-i-ty simply means: *a break or interruption in the normal physical structure of an article.*

Many discontinuities will not be half as long as the name given them.

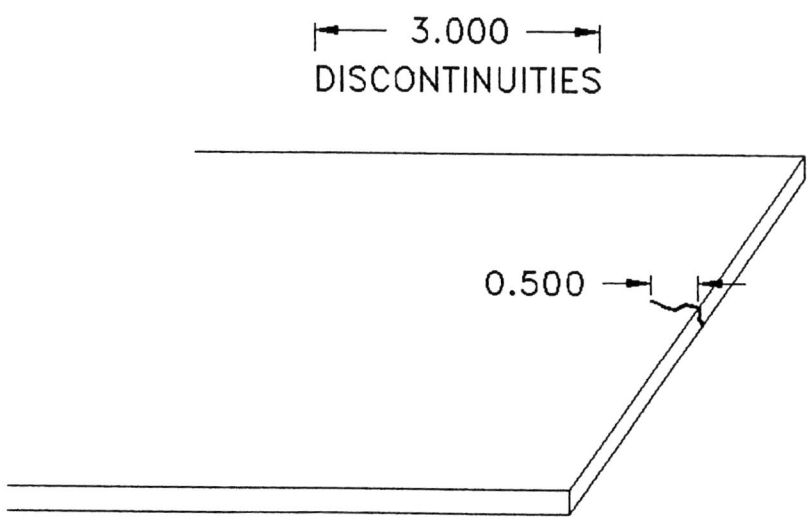

Of course, we are making a joke in the illustration; but in nondestructive testing, discontinuities are no joke. THEY ARE INTERRUPTIONS IN THE NORMAL PHYSICAL STRUCTURE OF AN ARTICLE.

Turn to the next page.

From page 1-19

1-20

A discontinuity in metal may be a hole, a crack, a flaw, or anything else that breaks the continuity of the metal. Discontinuities may be found on the surface of the metal or within the metal itself.

The portion of the article you see below has continuity of structure - it has no cracks or other flaws.

However, turn the article over, and you can see that the other side is cracked. The crack is an interruption in the normal physical structure of the article.

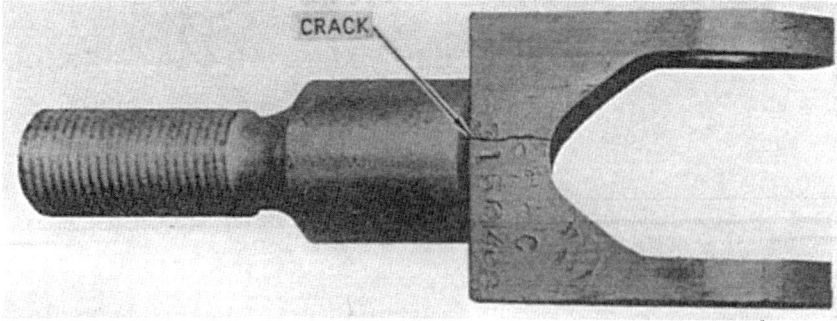

The crack in the above article would be called a:

discontinuity .... **Page 1-22**
discontinuation .... **Page 1-24**

From page 1-27

The cracks are discontinuities because they are *interruptions in the normal physical structure of the article*. A hole, a crack, or a flaw of any type is called a discontinuity.

The role of NONDESTRUCTIVE TESTS in the Quality Assurance Program then is to locate the many types of discontinuities found in the many types of materials used in end items such as the X-15, the space shuttle, space-launch vehicles and their associated spacecraft, ground-support systems, etc.

Turn ahead to page 1-23.

From page 1-20   1-22

Right. That crack is called a discontinuity - a break or interruption in the normal physical structure of the article.

Shown below are two launch-vehicle engine supports. Without visible discontinuities, they would look like this:

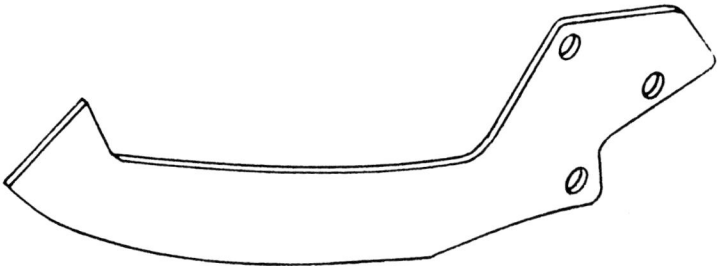

Which of the two below, if either, has a visible discontinuity?

1.

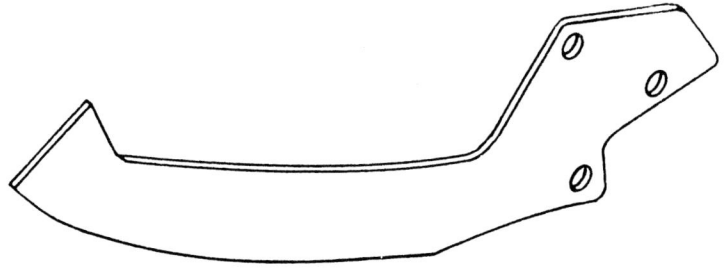

. . . Page 1-25

2.

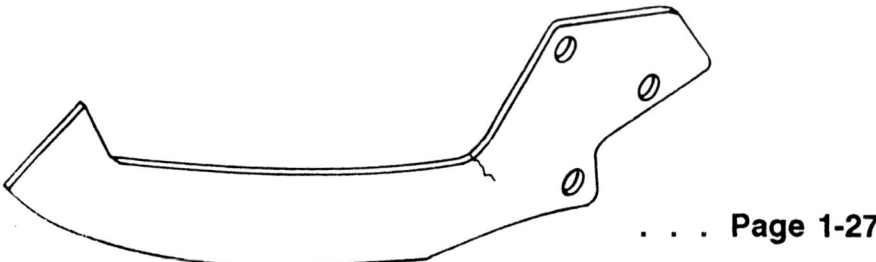

. . . Page 1-27

The Origins of Discontinuities

In the previous section, we pointed out the various nondestructive test methods and their role in the Quality Assurance Program. We found that each method provides visual indications of flaws, holes, cracks, etc., not normally visible to the unaided eye. We found that these breaks or interruptions in the normal physical structure of an article are called DISCONTINUITIES.

Before any nondestructive test method can be put to intelligent use, it is necessary to understand why discontinuities are found in materials. To do this, we are now going to discuss the refining processes that transform various mined ores into usable materials. We are also going to discuss the various metal-forming processes to understand why specific discontinuities take the shape they do. The smelting and forming processes are the guiding factors in determining the types of discontinuities and where they may be expected to appear in an article as a result of a specific forming process.

Since the causes for discontinuities in all metals are similar, we will discuss only the processing of steel in this chapter. Chapter 8 will introduce you to discontinuities in other engineering materials.

Some discontinuities found in ferrous metals have their beginning at the steel mill when the iron ore is melted in blast furnaces. Ferrous is defined as: *of or pertaining to iron*. Thus, steel is a combination of materials, most of which are derived from iron. Nonferrous metals include aluminum, brass, silver and bronze, among others.

Turn ahead to page 1-26.

No, that crack would not be called a discontinuation. This word is similar in meaning to the correct word but it is not the one used.

The crack in that article is to be called a *discontinuity*, which is defined as:

> A BREAK OR INTERRUPTION IN THE NORMAL PHYSICAL STRUCTURE OF AN ARTICLE.

Turn back to page 1-22 and continue.

From page 1-22   1-25

No, the first engine support did not have a visible discontinuity.

Remember, a *discontinuity is a break or interruption in the normal physical structure of an article.* Any kind of crack or flaw is called a discontinuity.

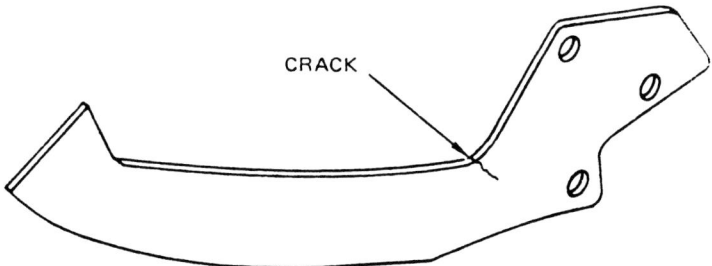

That crack in the engine support is the discontinuity. It is a break in the normal physical structure of the engine support.

Turn ahead to page 1-27 and continue.

Although more modern methods now exist, the blast furnace shown below simplifies the steel-making process. The steel-making process begins with iron ore, coke and limestone, which are fed into the top of a blast furnace. As the coke burns, an intense heat is created, which removes the oxygen from the iron ore and allows the molten metal to trickle to the bottom of the furnace.

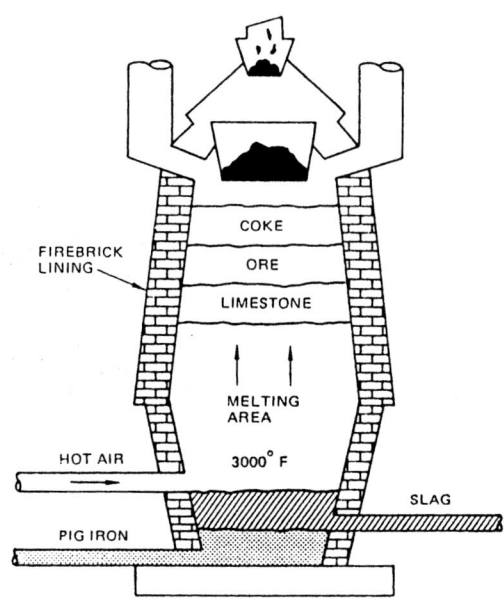

The limestone gathers the impurities in the iron ore to it, and the impurities become liquid. Like the iron, these impurities trickle to the bottom of the furnace; but, because they are lighter than the molten iron, the impurities remain on top and are called slag. Since the slag is made up of impurities, it is not wanted and is drawn out of the furnace. Most of the slag is removed in this way, but some remains and combines with the molten iron. This slag later forms some of the discontinuities found in metals.

Turn ahead to page 1-28 and continue.

From page 1-22

Fine. The second launch-vehicle engine support has a small but visible discontinuity - a crack.

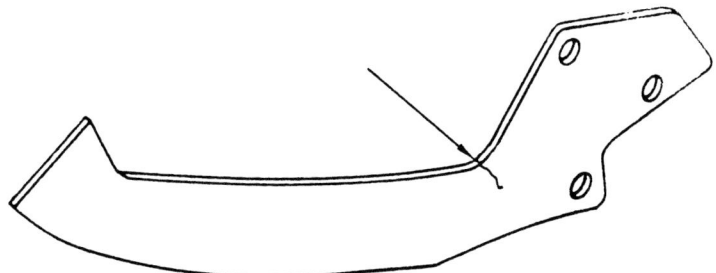

Here is an article that was to be used in a space-vehicle launcher. To the unaided eye, the cracks were not visible. Fortunately, the article was subjected to a nondestructive test, and the cracks were located before the article was installed in the launcher.

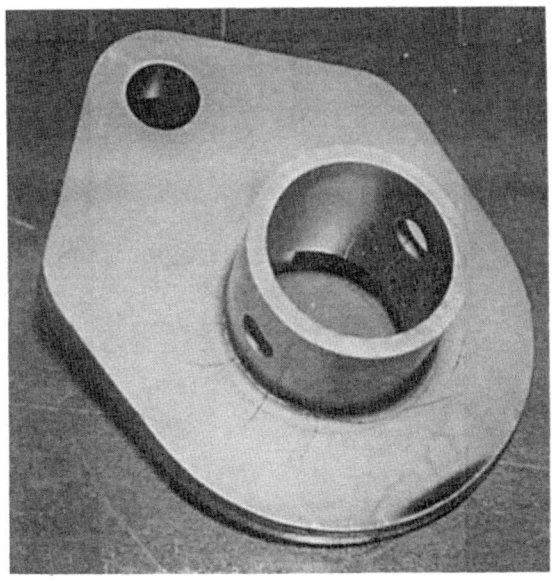

Turn back to page 1-21.

As the molten iron is drawn from the blast furnace, it is poured into molds to form what is called "pig iron." This name came from an early process in which molten iron was poured into large molds called "sows." These sows allowed the molten metal to trickle into smaller molds resembling suckling pigs; thus, the name pig iron.

Like its namesake, pig iron is not too clean. As the pig iron hardens, the slag impurities also harden into slate-like pockets within the metal. These pockets come from chemicals present in the original smelt that group together and linings within the furnace that introduce other impurities into the molten metal. The impurities are *not metallic* and were *accidentally included* in the iron. These pockets of slag in the iron are called NONMETALLIC INCLUSIONS.

Turn to the next page.

From page 1-28

Pig iron is the first product in the steel-making process. It is too brittle for most purposes, so it is further refined by one of the following processes: cupola, air furnace, electric process, bessemer process, puddling process, and the more common open-hearth furnace. The refining process remelts the pig iron, along with other materials, to make better quality metal. The products of these processes produce wrought iron, gray and white cast iron, nodular iron, and steel.

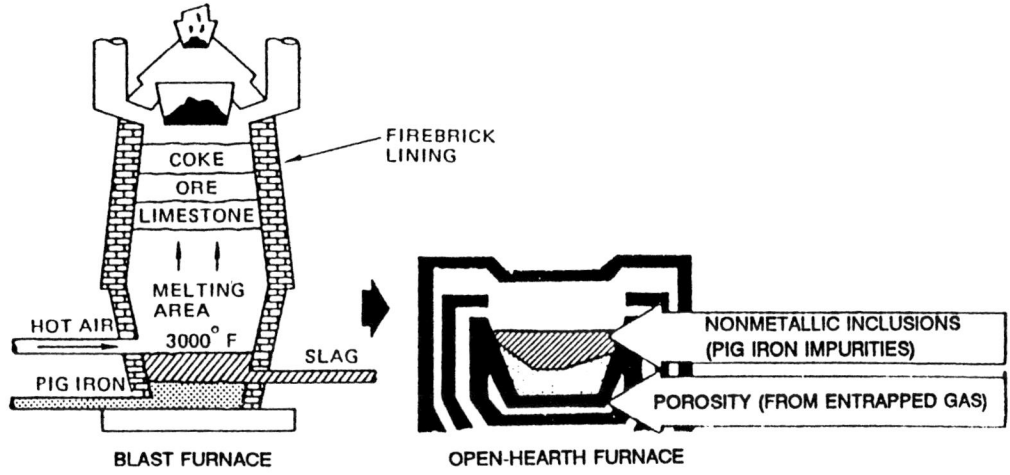

Pig iron is refined, and other metals or alloys are added to produce steel having desired physical properties.

Do you think the finished steel will still contain nonmetallic inclusions? Take a guess.

Yes .... Page 1-30
No .... Page 1-32

From page 1-29

Absolutely. Finished steel will contain NONMETALLIC INCLUSIONS. That was a good guess.

When the molten steel with its pig iron and other alloys is ready, it will be poured into molds to form INGOTS. Ingots may be small or they may weigh as much as 250 tons.

Pouring of the ingots is what we have been working toward, for metal parts are formed from the ingots.

Turn to the next page.

From page 1-30

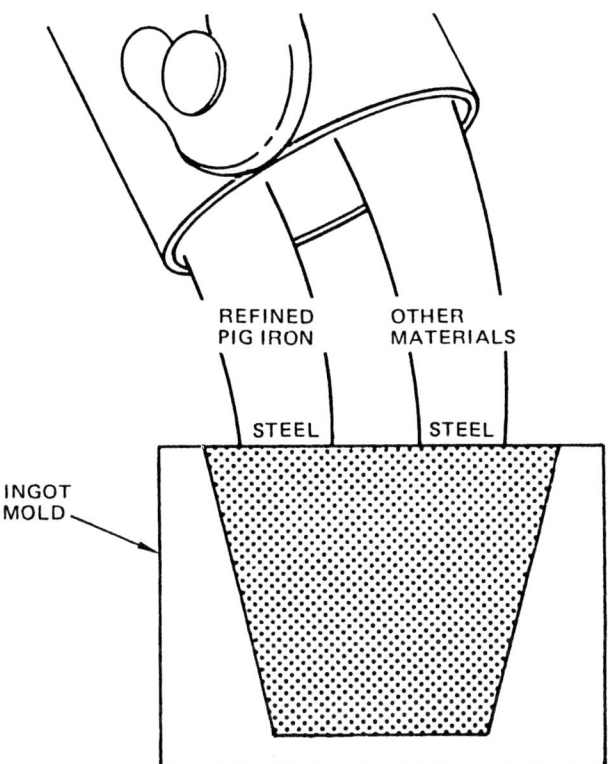

Here you see the molten steel as it is being poured into the ingot mold. Although the open-hearth-furnace refining process has eliminated some impurities, others have formed.

Would you expect to find nonmetallic inclusions in the above-pictured ingot?

Yes .... Page 1-33
No .... Page 1-34

You guessed "No." Actually, the finished steel will still contain nonmetallic inclusions, and here's why.

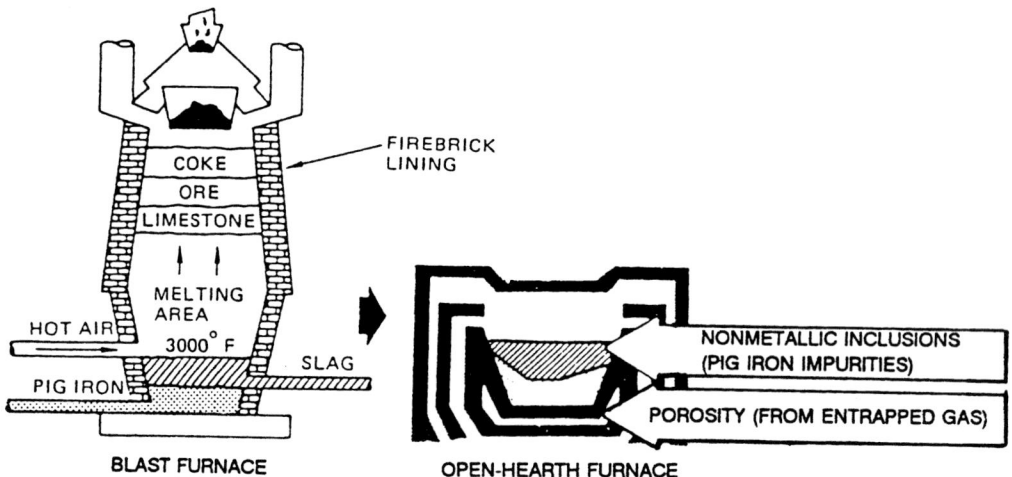

Most of the slag is eliminated at the blast furnace, but some of the impurities still remain with the molten iron. The molten iron is then placed in the open-hearth furnace where it is further refined; however, some of the nonmetallic material still remains. In addition, due to the process itself, gas bubbles, or porous areas, may form. So you see, the ingot has nonmetallic inclusions and, in addition, may have a certain amount of entrapped gas.

Turn back to page 1-30.

From page 1-31

That's right. You certainly would find nonmetallic inclusions in the ingot. Impurities formed as a result of the refining process cause pockets of foreign materials called nonmetallic inclusions.

The nonmetallic inclusions within the molten metal are lighter than the metal and rise toward the surface. Most of the nonmetallic inclusions manage to rise to the upper part of the ingot, but some are trapped. They did not have time to reach the surface before the metal hardened above them.

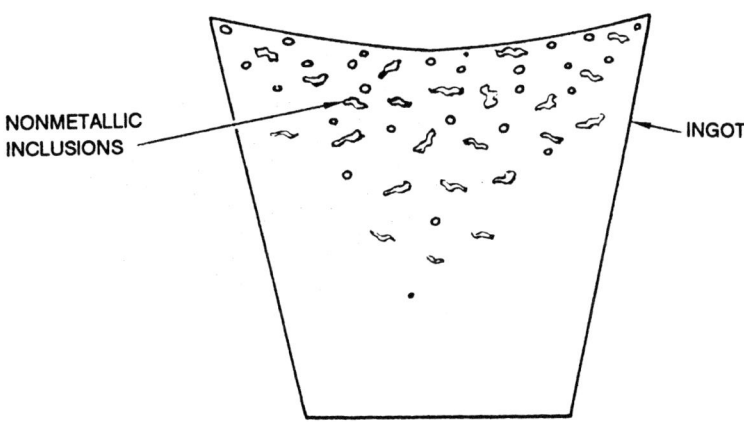

From the above, you can see that NONMETALLIC INCLUSIONS in the hardened ingot would look like which of the following?

Irregular discontinuities . . . . **Page 1-35**
Round discontinuities . . . . **Page 1-37**
Both of the above . . . . **Page 1-39**

From page 1-31   1-34

You selected "No." You would not expect to find nonmetallic inclusions in the ingot. Actually, the open-hearth furnace will eliminate some of the impurities or nonmetallic inclusions, but not all of them.

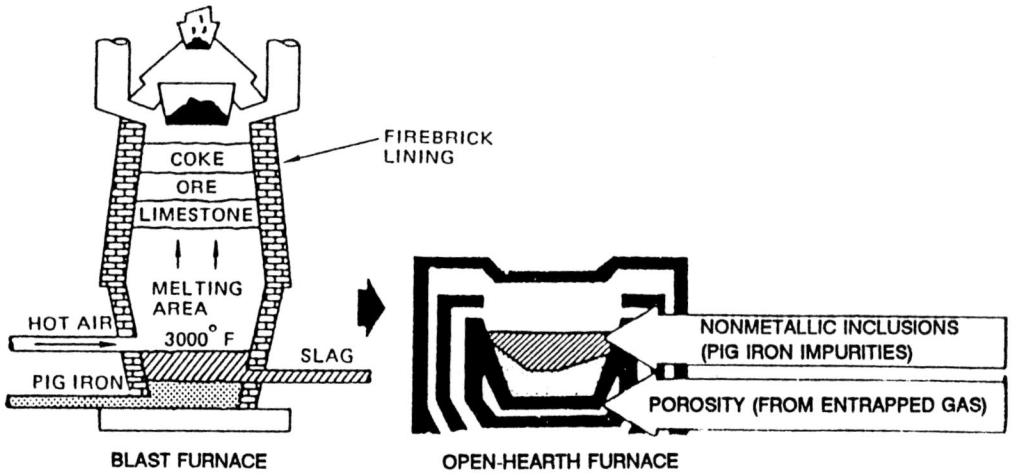

Remember that, although the refining process eliminates *some* of the impurities, it does not eliminate all of them. The ingot will contain some nonmetallic inclusions. In fact, as you can see from the above illustrations, other discontinuities are added.

Turn back to page 1-33.

From page 1-33

Right. Nonmetallic inclusions in ingots appear as irregularly-shaped discontinuities.

The irregularly-shaped, nonmetallic inclusions are not the only discontinuities found in ingots. Ingots also contain entrapped gas with a bubble-like appearance. These bubbles occur from gases formed in the metal when it was melted in the furnace. Some of the gas remains in the metal when it is poured into the ingot mold.

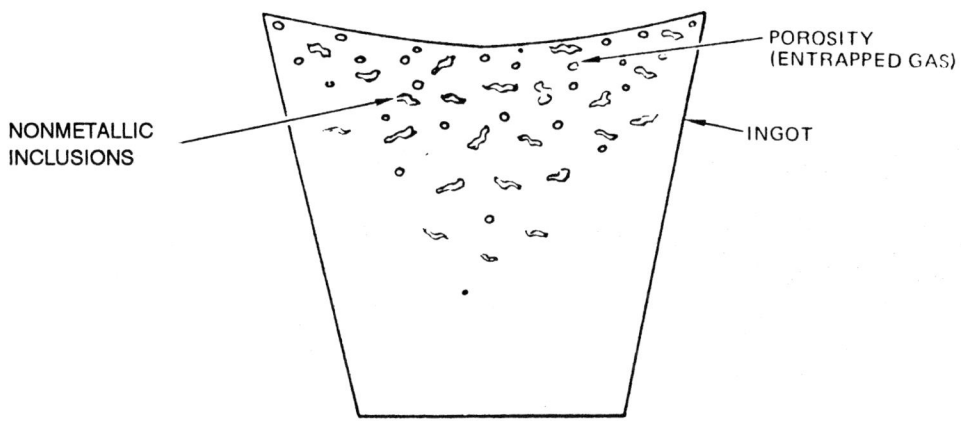

The gas bubbles are much like the bubbles in a bottle of pop. Like nonmetallic inclusions, the gas bubbles rise toward the surface as the ingot hardens. Most of the bubbles reach the surface. But again, as in the case of the nonmetallic inclusions, some of the bubbles are trapped by the hardening metal. They are called POROSITY.

From the illustration above, you can tell that POROSITY differs from nonmetallic inclusions because:

porosity is round in shape . . . . **Page 1-36**
porosity consists of round inclusions . . . . **Page 1-38**

Absolutely! Porosity is round.

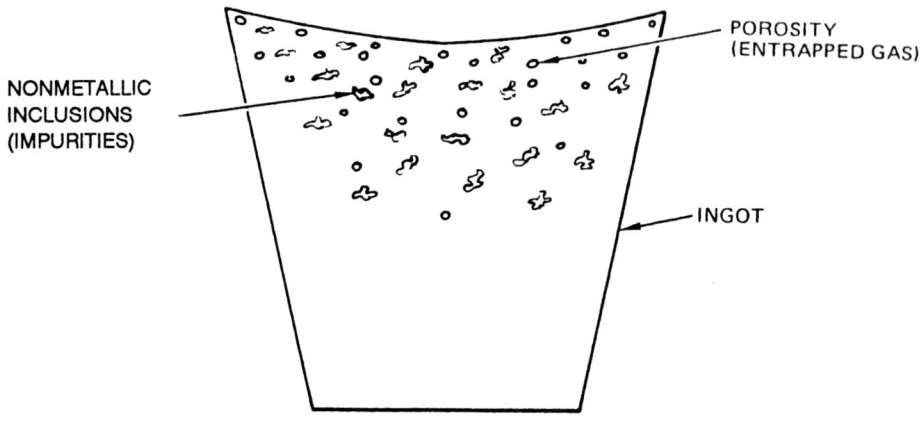

Since both nonmetallic inclusions and porosity are breaks or interruptions in normal physical structure of the ingot, they can be called:

breaks .... Page 1-40
flaws .... Page 1-42
discontinuities .... Page 1-44

You are guessing. Let's look at the ingot again. Those nonmetallic inclusions don't look round do they?

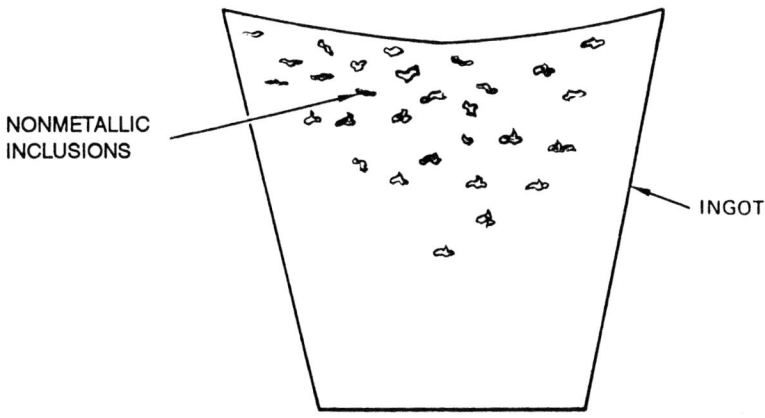

The dark spots in the ingot represent nonmetallic inclusions.

Turn back to page 1-33 and try again.

From page 1-35

You think porosity consists of round inclusions. No, you see porosity is formed by gas bubbles. Nonmetallic inclusions are formed by slag impurities.

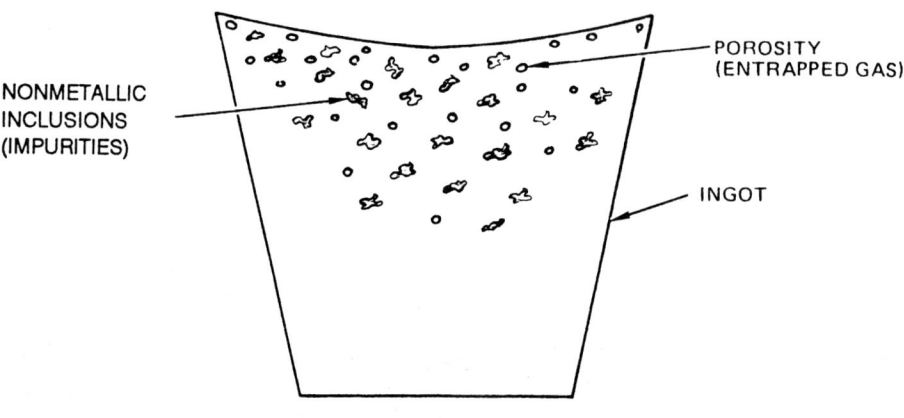

Here you can see the difference. Nonmetallic inclusions are pockets filled with foreign material. Porosity is caused by trapped gases, and the round holes have no solid material in them. So porosity is round, or nearly round.

Turn back to page 1-36.

You think that nonmetallic inclusions in a hardened ingot would appear both irregular and round in shape. Well, it is possible that you could find impurities that would approximate roundness, but it would most likely be very ragged.

In the usual sense, nonmetallic inclusions are very ragged. The dark spots in the illustration represent nonmetallic inclusions.

Turn back to page 1-33 and try again.

Yes, both porosity and nonmetallic inclusions are breaks in the structure of the ingot, but didn't we agree that all of the different names would be combined into one? We also gave the term we are to use a definition. Here it is:

> A BREAK OR INTERRUPTION IN THE NORMAL PHYSICAL STRUCTURE OF AN ARTICLE.

Turn back to page 1-36 and pick the correct name for this definition.

From page 1-44

In the molten condition, the metal expands. In other words, molten metal occupies more space than does the hardened metal. When the molten metal starts to cool and harden, it shrinks. Since the center of the ingot is last to cool and harden, most of the shrinkage is absorbed in the center. This results in the cavity called "PIPE." Note that the ingot is larger at the top, which extends the time the metal stays liquid and minimizes the shrinkage. Pipe may be deep into the ingot and layered or "bridged," which will most likely remain in the finished product.

Since "pipe" in the end of the ingot makes that portion of the ingot unusable, most ingot molds in use today have an added portion that acts as a reservoir and absorbs most of the ingot shrinkage. The reservoir is called the "hot top."

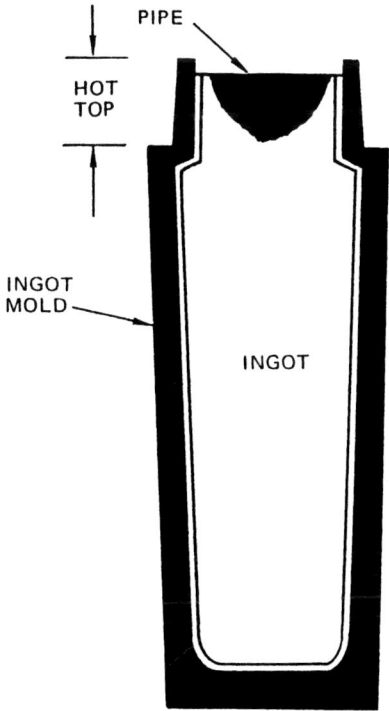

Turn ahead to page 1-43.

From page 1-36

Yes, both porosity and nonmetallic inclusions are flaws in the structure of the ingot. But didn't we agree that all of the various names would be combined into one for our purposes? We also gave the term we are to use a definition. Here is the definition:

> A BREAK OR INTERRUPTION IN THE NORMAL PHYSICAL STRUCTURE OF AN ARTICLE.

Turn back to page 1-36 and select the word we have defined.

The "hot top" is designed to absorb as much of the pipe as possible. It is also used to absorb some of the nonmetallic inclusions and porosity. As we have mentioned before, the type and quantity of a specific discontinuity are dependent upon the method of metal processing and on the type of ingot mold.

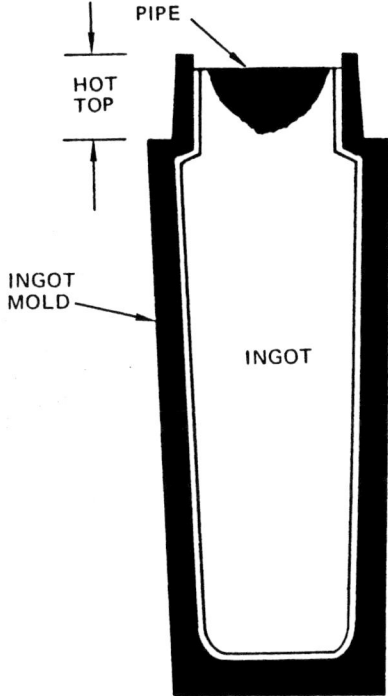

Do you think the "hot top" would eliminate all of the nonmetallic inclusions, gas-bubble porosity, and pipe?

Yes .... **Page 1-45**
No .... **Page 1-47**

From page 1-36   1-44

Very good. Both nonmetallic inclusions and porosity are breaks or INTERRUPTIONS IN THE NORMAL PHYSICAL STRUCTURE OF THE INGOT and are, therefore, DISCONTINUITIES.

Ingots are made in many sizes and shapes. The shape depends upon how the ingot is to be used. If the ingot is to be rolled into plate stock, it would be poured in a relatively flat, rectangular mold. If the ingot is to be used for making bar stock, it would be poured into a longer and thinner, round or square mold. The different types of ingot molds, in conjunction with the different steel-making methods, result in other types of discontinuities. Remember, the ingot is the refined pig iron, or steel, in our example. Let us consider one of these now.

Here is an ingot designed to be rolled into bar stock. A severe shrinkage depression is visible at the top; it is caused by molten metal shrinking when it cools and solidifies. The depression is called "PIPE."

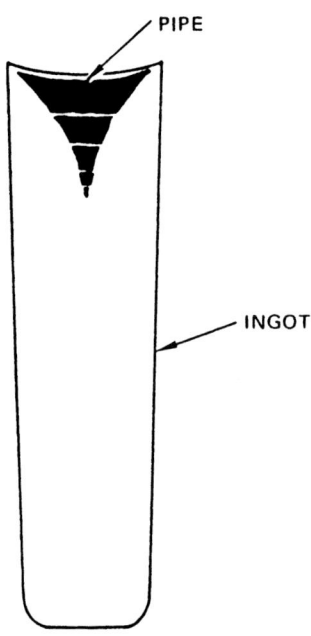

Turn back to page 1-41.

You think the hot top *would* eliminate all of the nonmetallic inclusions, gas-bubble porosity, and pipe. Actually, those discontinuities are more widely distributed than was indicated in that last illustration. Take a look at this one.

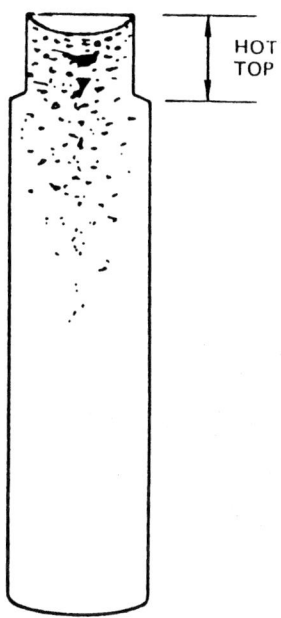

Here you can see the distribution of discontinuities in this particular ingot, and they are not all located in the "hot top." Of course all ingots will not have such a wide distribution of discontinuities as this one. The point to be made here is that the "hot top" does not absorb all of the discontinuities; therefore, some of the discontinuities in the ingot will still be present below the hot top.

Turn ahead to page 1-47 and continue.

From page 1-51   1-46

Yes, of course. You would expect to find porosity in the ingot. Entrapped gas causes porosity. Nonmetallic inclusions may also be present.

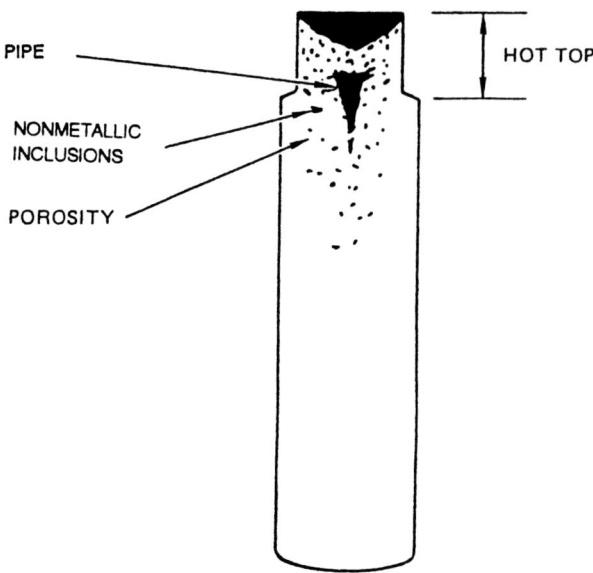

In summary, there are three main types of discontinuities in the ingot:

- *Porosity*, which is round, or nearly round, and which is caused by entrapped gas in the molten metal.

- *Nonmetallic inclusions*, which are irregularly shaped and consist of slate-like impurities accidentally included in the molten metal.

- *Pipe*, caused by shrinkage as the molten metal solidifies. It may extend deep into the center of the ingot.

Turn ahead to page 1-50 and continue.

Correct. The hot top will not eliminate all of the different discontinuities.

Here are some ingots cast with and without hot tops. The ingot on the left was poured in a mold without a hot top, and you can see the pipe (shrinkage cavity) extends deep into the center of the ingot. The ingot on the right was poured with a hot top, and you see the pipe still extends very deep into the ingot.

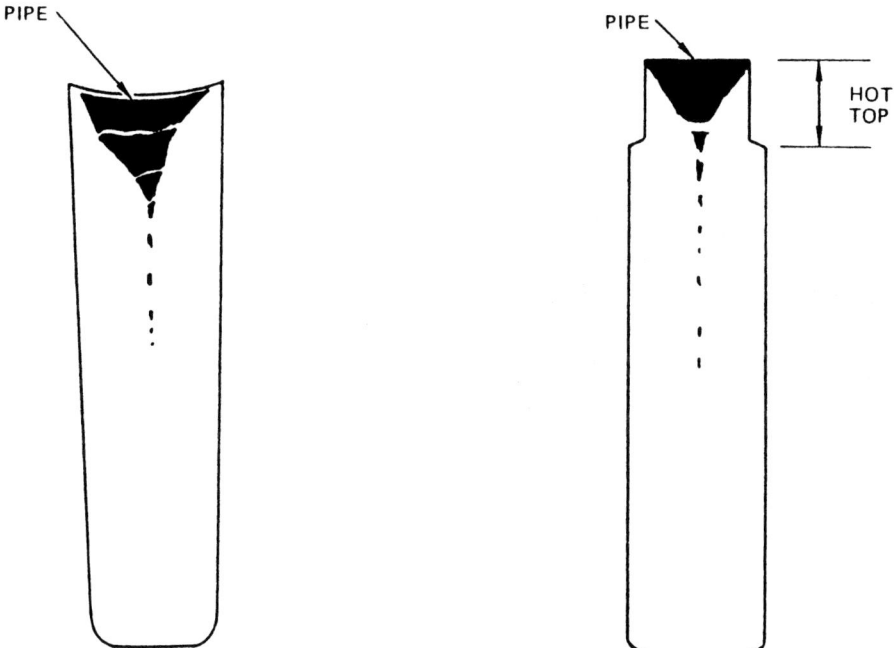

The only difference here is the shape of the shrinkage cavity or "pipe" in the two ingots. The intermittent cavities extending deep into the ingot are individual extensions of the main cavity. They are also called pipe, as they were caused by shrinkage as the molten metal solidified.

We can conclude from the above that pipe is caused by:

expansion as the metal solidifies .... **Page 1-49**
shrinkage as the metal solidifies .... **Page 1-51**

From page 1-51

Yes, you would probably find nonmetallic inclusions in that ingot. But the point we were trying to emphasize was that there was excessive entrapped gas. In fact, the gas pressure was great enough to force molten metal through the hardening top of the ingot.

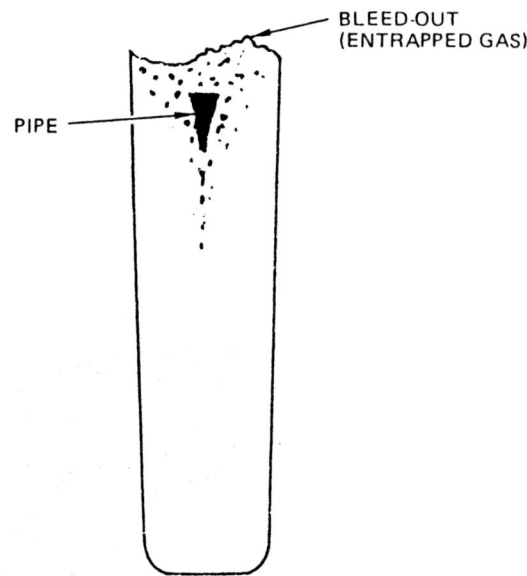

Where there is gas, there are gas bubbles that cause porosity.

Turn back to page 1-46.

You think that "pipe" is caused by expansion as the metal solidifies. No, the metal does not expand when it starts to harden. It shrinks. These two illustrations demonstrate our point.

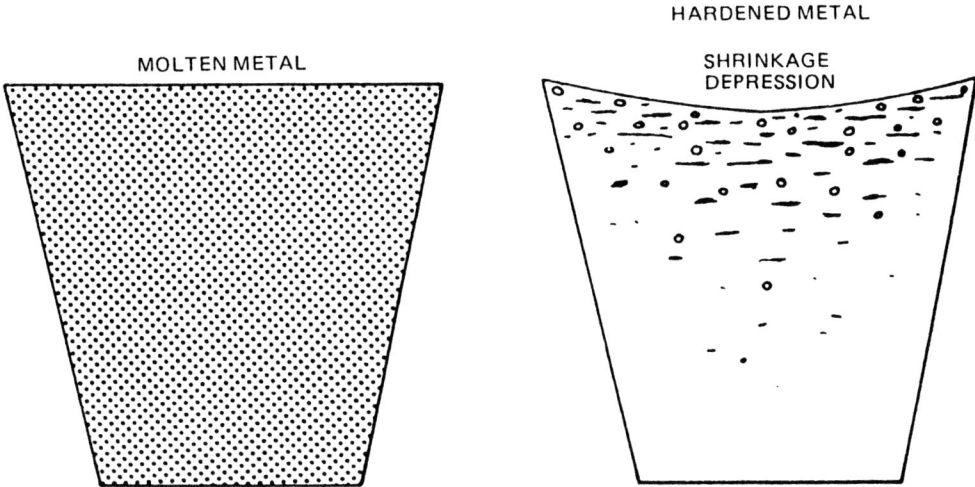

When the molten metal is poured into the mold, it fills the mold. When the metal hardens it shrinks, causing a shrinkage depression as shown by the illustration on the right. In the ingot we were discussing, the shrinkage was more pronounced. This resulted in the large cavity and the intermittent cavities called "pipe." So you see, "pipe" is caused by shrinkage rather than expansion of the metal as it solidifies.

Turn ahead to page 1-51.

From page 1-46   1-50

When the ingot solidifies with many of the discontinuities contained in the upper portion, the "hot top" is cut off. To use a steelman's term, the top of the ingot is "cropped."

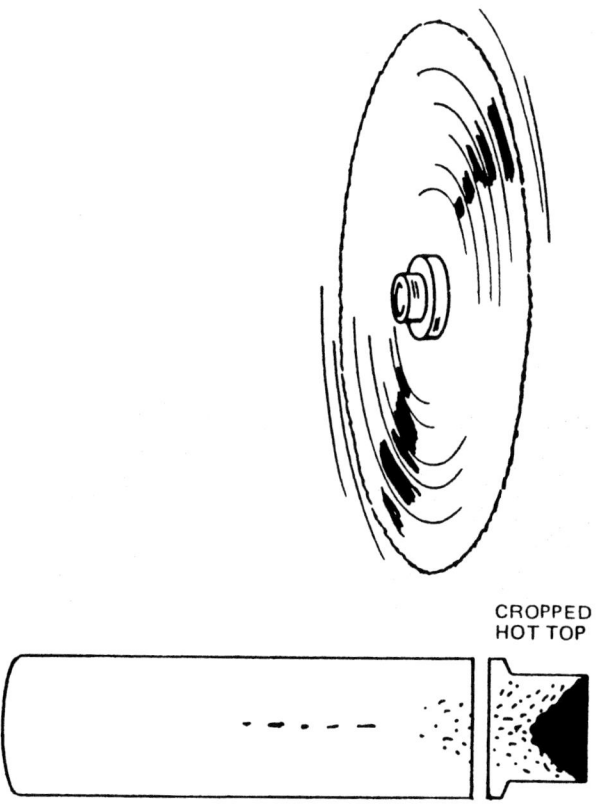

CROPPED HOT TOP

Would cropping the top of the ingot remove many discontinuities?

Yes .... **Page 1-53**
No .... **Page 1-54**

From page 1-47   1-51

Absolutely. Pipe is caused by shrinkage as the metal solidifies.

Pipe is not always obvious in an ingot. It can be covered up. Here is an example.

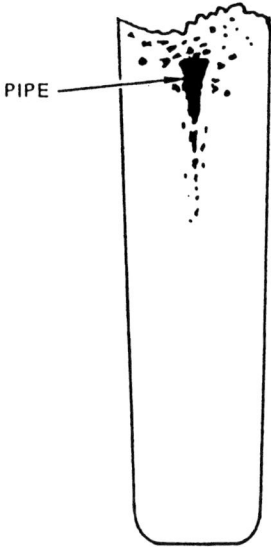

In this case, excessive gas pressure forced molten metal up through the hardening top causing a bleeding action. The bleeding covers the pipe as shown in the illustration.

Since excessive gas pressure was present in this ingot, what other type of discontinuity would you expect to find in the ingot?

Porosity . . . . **Page 1-46**
Nonmetallic inclusions . . . . **Page 1-48**

From page 1-53

You think all discontinuities will be eliminated by cropping. That is not right. Unfortunately, in most cases not all of the porosity, nonmetallic inclusions, and pipe will be eliminated. Consider the deep-lying shrinkage cavities in this ingot.

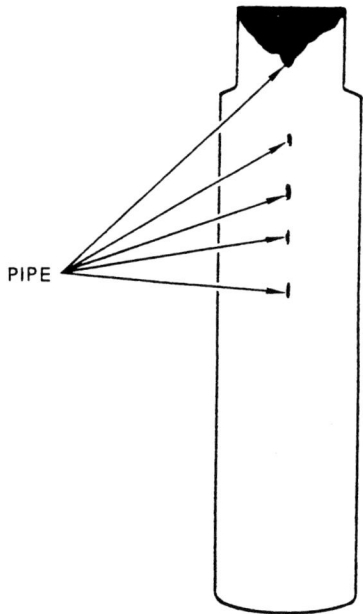

While this ingot is relatively clean, four shrinkage cavities or "pipe" would be left after the top of the ingot is cut off. So you see, not all of the discontinuities will be eliminated by cropping the ingot.

Turn ahead to page 1-55.

From page 1-50   1-53

Very good. The cropping or removal of the "hot top" would remove most of the nonmetallic inclusions and porosity. It would also eliminate most of the shrinkage depression or "pipe."

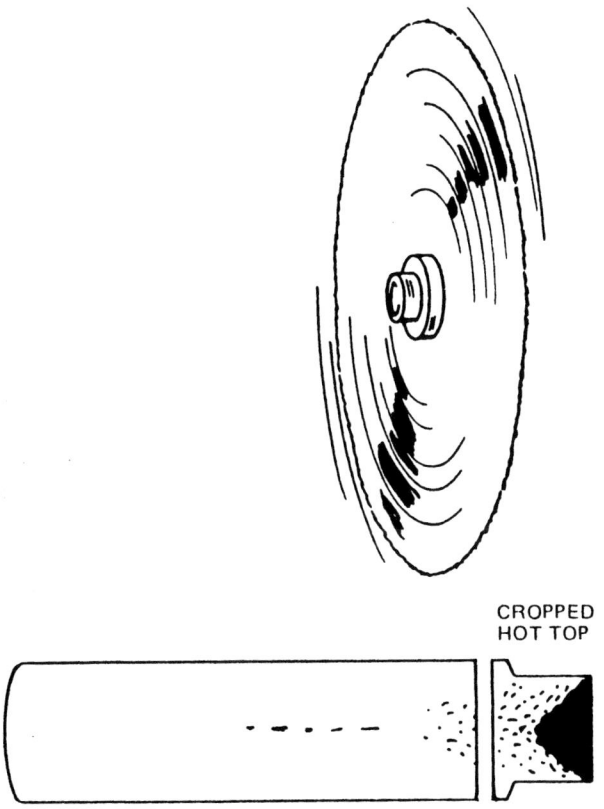

CROPPED HOT TOP

Cropping will eliminate all of the "pipe" in the ingot except in those cases where intermittent shrinkage cavities extend deep in the ingot. But are all discontinuities eliminated by cropping?

Yes .... **Page 1-52**
No .... **Page 1-55**

From page 1-50

You selected "No." You must have misread the question. Remember, CROPPING means to cut off or remove the "hot top." The hot top is put there to accumulate as many of the discontinuities as possible.

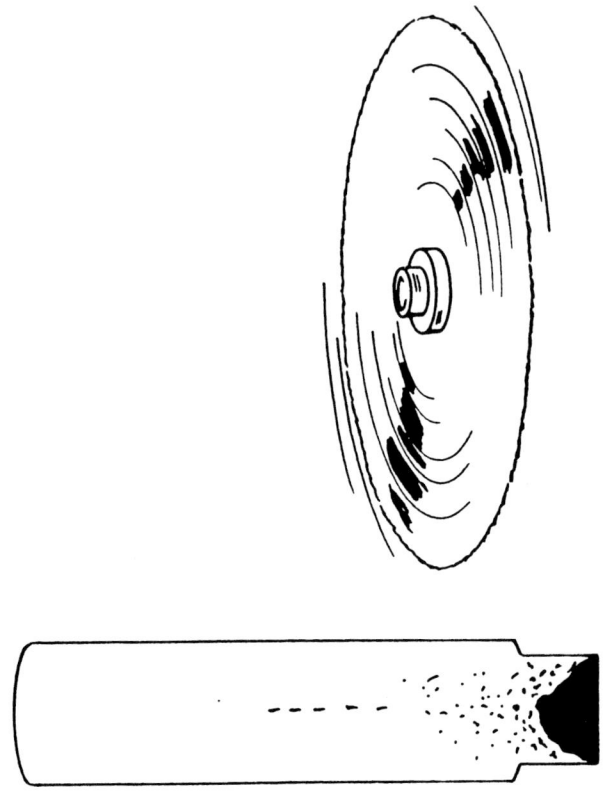

Cutting the top off the ingot will not remove all of the discontinuities, but it will eliminate many of them. Here again, remember that the number and type of discontinuities in a specific ingot will depend upon the type of mold used and the smelting process. But cropping the ingot will usually remove many of the discontinuities.

Turn back to page 1-50 and try again.

Right. Not all of the discontinuities are eliminated by cropping the ingot. Most of them are eliminated by removing the "hot top." But a few discontinuities remain in the main portion of the ingot.

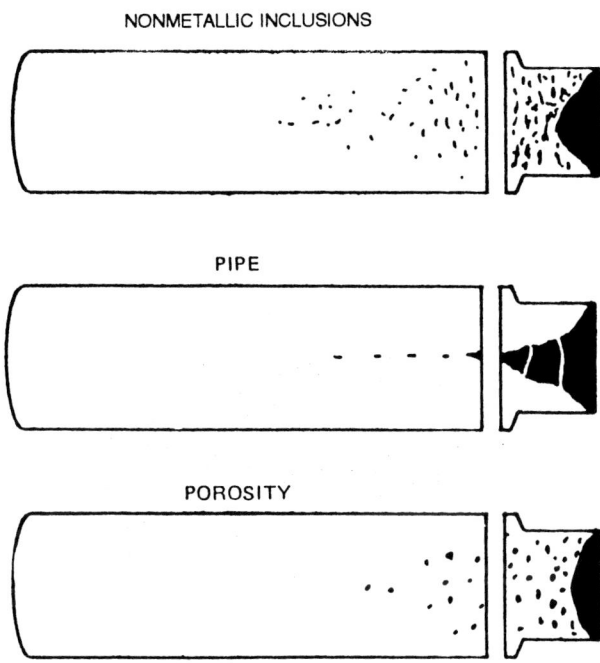

All of these discontinuities may remain in a single ingot, depending upon the type of ingot mold and the method used in refining the metal.

After the top is cut off, the ingot gets a new name - BLOOM - which in turn can be processed into smaller size SLABS or BILLETS. The slab or billet is normally the starting point for actual forming of articles or materials.

Let's briefly review the material we have covered before we focus on metal-forming techniques.

Turn to the next page.

From page 1-55

CHAPTER REVIEW

The next few pages are different from the ones which you have been reading. At the end of each chapter there is a Chapter Review. Following each question you should choose the letter that <u>best</u> describes the correct response.

After completing ALL of these Chapter Review questions, you will find the answers at the end of the chapter. A page number reference is also listed if you decide to review the subject matter.

Now for the review.

_____ 1. Nondestructive tests are methods of detecting interruptions in the normal or physical structure of:

      A.   a discontinuity.
      B.   an article.
      C.   a flaw.
      D.   an ingot.

_____ 2. Any crack, break, or flaw in the continuity of the physical structure of an article is called:

      A.   a break.
      B.   pipe.
      C.   a discontinuity.
      D.   shrinkage.

3. Many discontinuities have their origin in the refining process and are in the molten metal poured into the _____ mold.

   A. ingot
   B. pipe
   C. nonmetallic
   D. furnace

4. Slag impurities are not entirely in the refining process and are found in the ingot as foreign material called _____ inclusions.

   A. nonmetallic
   B. pipe
   C. furnace
   D. porosity

5. Nonmetallic inclusions are ragged when they harden and take _____ shape.

   A. a round
   B. a regular
   C. an undetermined
   D. an irregular

From page 1-57

____ 6. Due to the refining process itself, all of the material-forming gas is not removed from the molten metal. This causes entrapped gas bubbles to form which are called:

    A. pipe.
    B. porosity.
    C. inclusions.
    D. irregularity.

____ 7. Gas-bubble porosity in the ingot is like bubbles in a bottle of pop. When the ingot hardens, the porosity takes on the shape of the gas bubbles that are _____ or nearly _____.

    A. round, irregular.
    B. irregular, irregular.
    C. square, square.
    D. round, round.

____ 8. Another discontinuity is formed as the molten metal shrinks as it hardens. As a result, the top of the ingot has:

    A. a protrusion.
    B. shrinkage.
    C. blanks.
    D. articles.

_____ 9. Due to the shape of some ingot molds, the shrinkage depression can be severe, extending deep into the center of the ingot. This type of depression is called:

A. protrusion.
B. expansion.
C. swelling.
D. pipe.

_____ 10. Most ingot molds have reservoirs called a "hot top" to absorb as much of the pipe, porosity, and nonmetallic _____ as possible.

A. protrusions
B. shrinkages
C. irregularities
D. inclusions

_____ 11. Pipe, porosity, and nonmetallic inclusions are all interruptions in the normal physical structure of the ingot and are therefore:

A. metallic inclusions.
B. shrinkages.
C. discontinuities.
D. desirable.

From page 1-59

_____ 12. Most discontinuities are removed from the ingot when the "hot top" is:

A. cut off or cropped.
B. cooled down.
C. replaced.
D. molded.

Turn to the next page for answers to these review questions.

ANSWERS TO REVIEW QUESTIONS FOR CHAPTER 1

Question & Answer		Reference Page(s)
1.	B	1-13
2.	C	1-19
3.	A	1-33
4.	A	1-28
5.	D	1-35
6.	B	1-35
7.	D	1-38
8.	B	1-44
9.	D	1-44
10.	D	1-43
11.	C	1-46
12.	A	1-50

Turn to the next page and begin Chapter 2.

CHAPTER 2

WORKING THE BILLET

Now that you are familiar with steel making from the blast furnace to pig iron, we are going to discuss the working or forming of the remaining portion of the cropped ingot - a *slab* or *billet*. Recall that the ingot still contained discontinuities such as porosity, nonmetallic inclusions, and pipe. We then cropped the ingot to remove a large portion of these.

Working of the billet centers around a fundamental fact concerning the physical structure of the cast ingot.

> The cast ingot is made up of millions of crystals that form a CRYSTAL GRAIN STRUCTURE.

Therefore, the billet has a crystal grain structure. If a magnifying glass were strong enough, the magnified grains would look something like this.

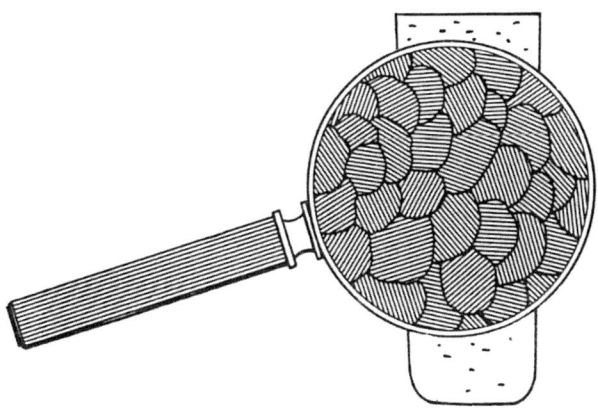

Turn to the next page.

"Working" the metal, such as rolling the billet between heavy rollers to a desired shape, breaks these crystals down. A finer grain is formed, and the crystals are STRETCHED OUT IN THE DIRECTION OF ROLL.

The billet is usually rolled after it has been heat-soaked in a furnace so that it is evenly heated throughout. Heating allows the crystals to break more easily into smaller grain-shaped crystals in the metal.

After heat-soaking to attain a uniform temperature, the billet is forced between large, heavy rollers. This rolling reduces the thickness of the billet and increases its length.

Turn to the next page.

From page 2-2   2-3

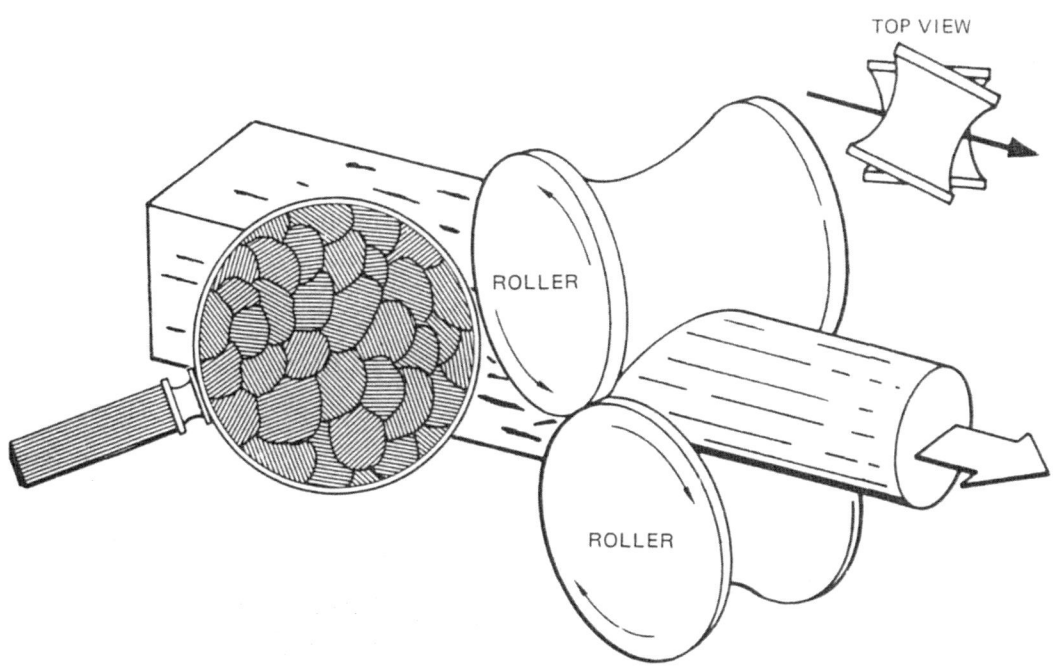

Here you can see the thickness of the billet has been reduced and the length increased. The pressing or squeezing action of the rollers has broken up the crystals which re-form into smaller crystals and a more even grain. The crystals are being stretched out IN THE DIRECTION OF ROLL.

If the billet is rolled another time to increase the length, what do you think will happen to the crystal structure?

**The crystals will break into smaller crystals and
the grain will be finer** .... Page 2-5
**The crystals will break up and spread out in all
directions** .... Page 2-7

From page 2-8   2-4

If the slab is to be rolled into sheet or plate material, it is rolled between wide, heavy rollers. These wide rollers reduce the slab thickness and increase its width and length.

With this in mind, how do you think a nonmetallic inclusion would act as it is pressed between the rollers?

**It would spread in all directions, but mainly in the
direction of roll .... Page 2-6
It would spread out evenly in all directions .... Page 2-9**

Right you are. The crystals will form finer grain. IT IS IMPORTANT TO REMEMBER THAT GRAIN IN THE METAL IS FORMED IN THE DIRECTION IN WHICH THE METAL IS ROLLED.

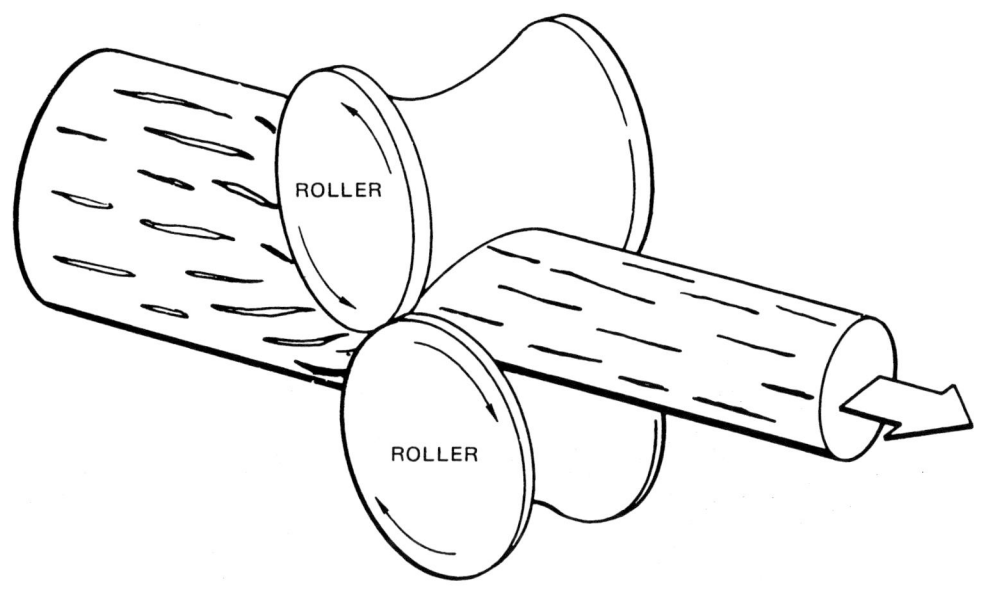

Steel is rolled to reduce it in size and to shape it as nearly as possible to the shape of the finished product. It is also rolled to refine the grain structure, which makes the metal much stronger.

Which way does the grain form in the bar being rolled above?

Through the length of the bar . . . . Page 2-8
Through the width of the bar . . . . Page 2-10

Right. A nonmetallic inclusion would spread out in all directions but mainly in the DIRECTION OF ROLL.

As the slab is rolled thinner, longer, and wider, the nonmetallic inclusion is flattened. This flattened inclusion is now called a LAMINATION.

A lamination is a nonmetallic inclusion that is sandwiched in the steel.

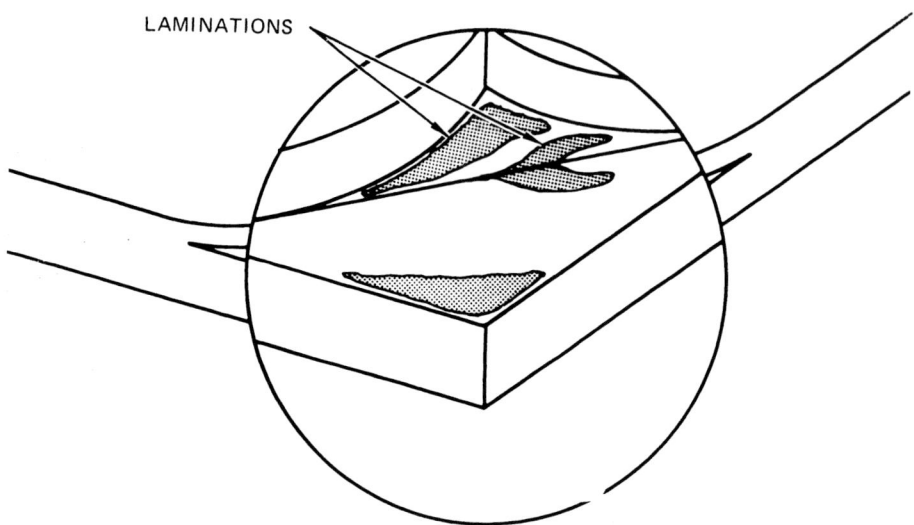

Here you can see the lamination in the flattened condition. Notice also that they have been stretched in the direction of roll in the same manner as the crystal grain structure.

What do you think would happen to porosity in the slab being rolled into sheet or plate material?

Porosity would cause a lamination .... **Page 2-11**
Porosity would disappear .... **Page 2-13**

From page 2-3

Your answer, ". . . crystals will break up and spread out in all directions," isn't the one we wanted. Let's see why.

The billet is heated to allow the crystals to more easily break into smaller crystals by the pressing or squeezing action of the rollers. As the crystals pass through the rollers, they break into smaller crystals with a more even grain.

There you have it. Before rolling, the crystal grain in a billet is irregular. After rolling, this grain is more regular and it has direction - the direction in which it was rolled. If this metal were now cut in half, the grain would be visible to the naked eye.

Turn back to page 2-5.

Good for you. The grain runs through the length of the bar. The grain will always run in the DIRECTION OF ROLLING.

Now let's go back to the billet or slab and see what effect rolling has on the nonmetallic inclusions. As you remember, nonmetallic inclusions are pockets of foreign materials trapped in the hardened ingot. Most of the nonmetallic inclusions are eliminated when the ingot is cropped, but some remain in the billet or slab.

Here is a slab being rolled, and the grain is formed in the direction of roll. The grain is formed when the crystals are broken and re-formed into small crystals which form the grain. Now let's see what happens to the nonmetallic inclusions during rolling.

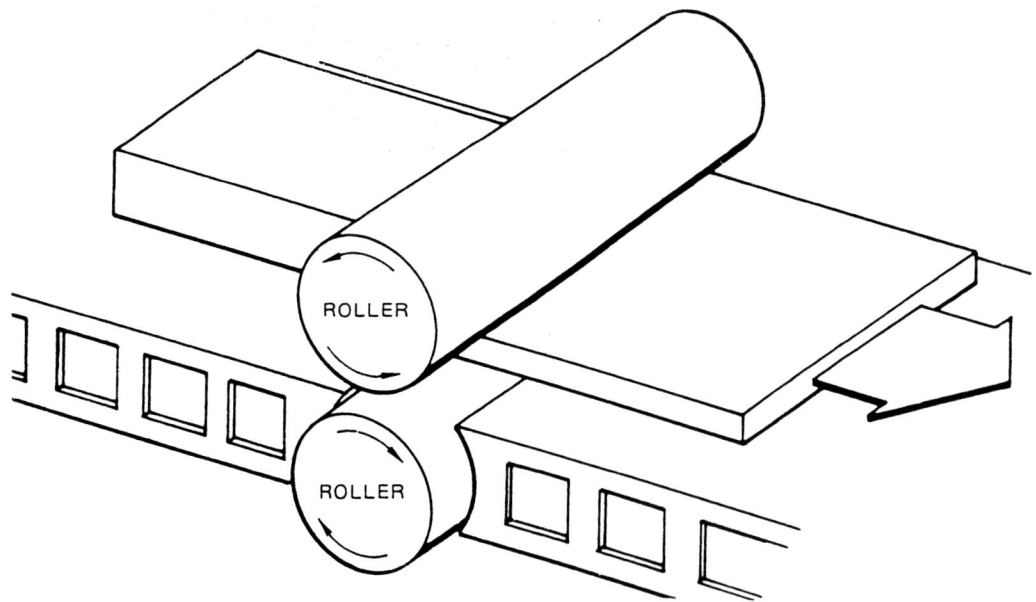

Turn back to page 2-4.

From page 2-4

In your opinion a nonmetallic inclusion in a slab being rolled into sheet or plate material would spread evenly in all directions. This is not the case. The inclusion spreads mainly in the direction of roll, but that is not the only direction in which it spreads. Let's see why.

When flattened into sheet or plate, a slab becomes (through repeated rolling) longer, wider, and thinner. It is being pressed forward much like pie dough under a rolling pin. The rolling flattens the inclusions in many directions, although mainly in the direction of roll.

Turn back to page 2-4, re-read the paragraph, and choose the correct answer.

From page 2-5   2-10

You think the grain runs through the width of that rolled bar. Actually the grain runs through the length of the rolled bar. Here is another illustration to make the point clear.

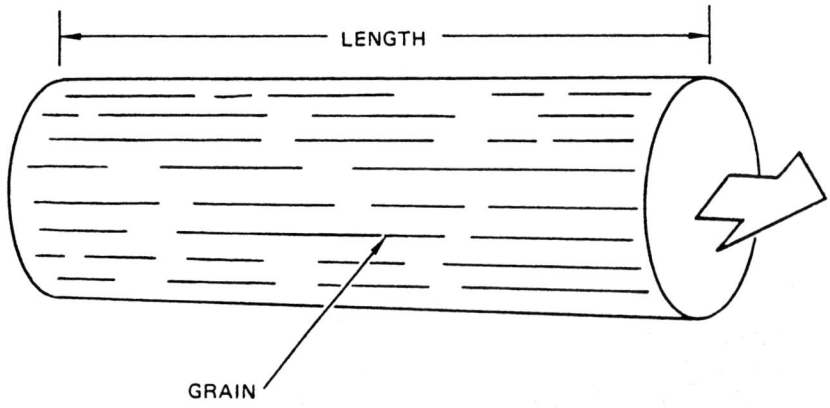

The width of the bar in this case is the diameter. Notice the lines which indicate the direction of grain. Doesn't the direction of the grain run the length of the bar?

Turn back to page 2-8.

Yes indeed! Porosity would cause a lamination. The gas porosity in the slab would spread out in all directions, but mainly in the DIRECTION OF ROLL. Porosity could possibly disappear in the rolling of some types of material, such as aluminum. This result would depend upon the temperature of the material and the method or amount of rolling. In this case the metal would simply fuse or join due to the temperature and pressure applied in the rolling process. If the metal does not fuse, the porosity would cause a lamination.

Here is an actual lamination caused by a nonmetallic inclusion. It was located in the center of a rolled piece of plate stock. The plate was sawed in half to expose the inclusion.

Turn to the next page.

You can see that the nonmetallic inclusion has been flattened and spread out, but MAINLY IN THE DIRECTION OF ROLL.

Pipe also causes lamination in rolled sheet or plate material. The squashing action of the rolling process flattens the pipe in the same way that porosity is flattened. Both types of discontinuities are flattened or spread out in all directions, but MAINLY IN THE DIRECTION OF ROLL.

Turn ahead to page 2-14.

From page 2-6

You think porosity would disappear in a slab rolled into sheet or plate material. Yes, that is true in some kinds of materials which fuse or weld easily under proper temperature and pressure conditions.

What we are concerned with here is the porosity that does not disappear. In this case, the round, bubble-like discontinuities are flattened when the slab is rolled. They spread out in all directions, but mainly in the direction of rolling. Because of this characteristic, porosity will cause a lamination in much the same way that a nonmetallic inclusion causes a lamination.

Turn back to page 2-11.

From page 2-12

We have seen what happens to large porosity, pipe, and nonmetallic inclusions in a slab being rolled into sheet or plate material. These discontinuities flatten and spread out in all directions, but mainly in the direction of roll. These flattened discontinuities are called LAMINATIONS. Now let us see how these discontinuities react when a billet is rolled into bar stock.

When the billet is rolled into bar stock, it is forced between rollers having an opening smaller than the size of the billet. The rollers have raised edges which are close together, preventing the billet from being squeezed to the sides under the pressure. The billet can only roll forward and emerge smaller in diameter and longer.

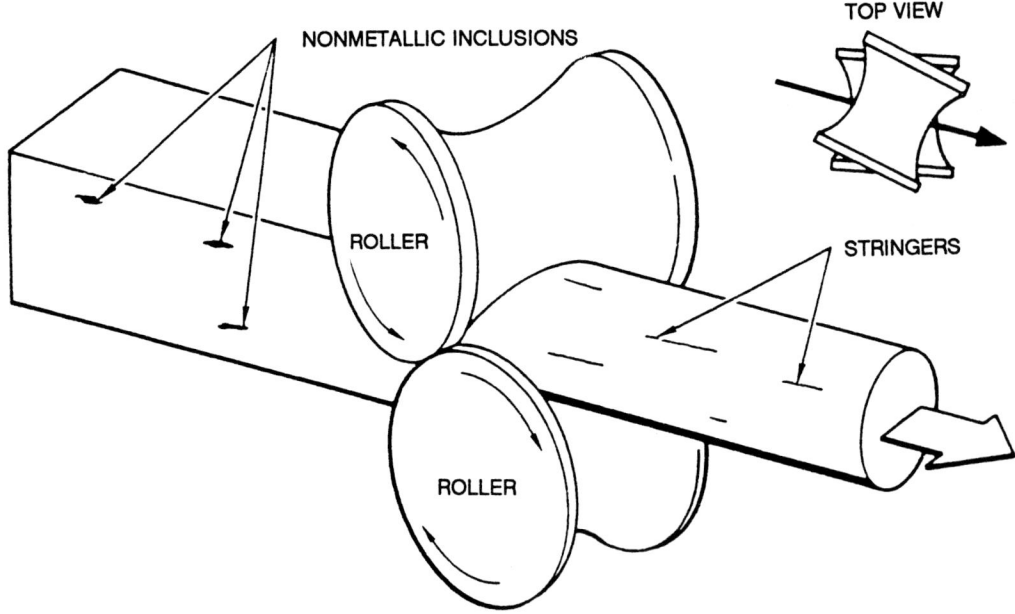

Turn to the next page.

From page 2-14

The longer the billet becomes, the longer and thinner the nonmetallic inclusion becomes. The stretched-out inclusion is now called a STRINGER as illustrated in the figure on the facing page.

Which of the following best describes how a nonmetallic inclusion reacts in a billet being rolled into bar stock?

It spreads out like an egg in a frying pan .... **Page 2-16**
It extends in the direction of grain formation .... **Page 2-19**

No, a nonmetallic inclusion in bar stock does not spread out like an egg in a frying pan. You are thinking of a lamination - an inclusion in sheet or plate material that spreads in many directions, but mainly in the direction of rolling. Here we are concerned with bar stock.

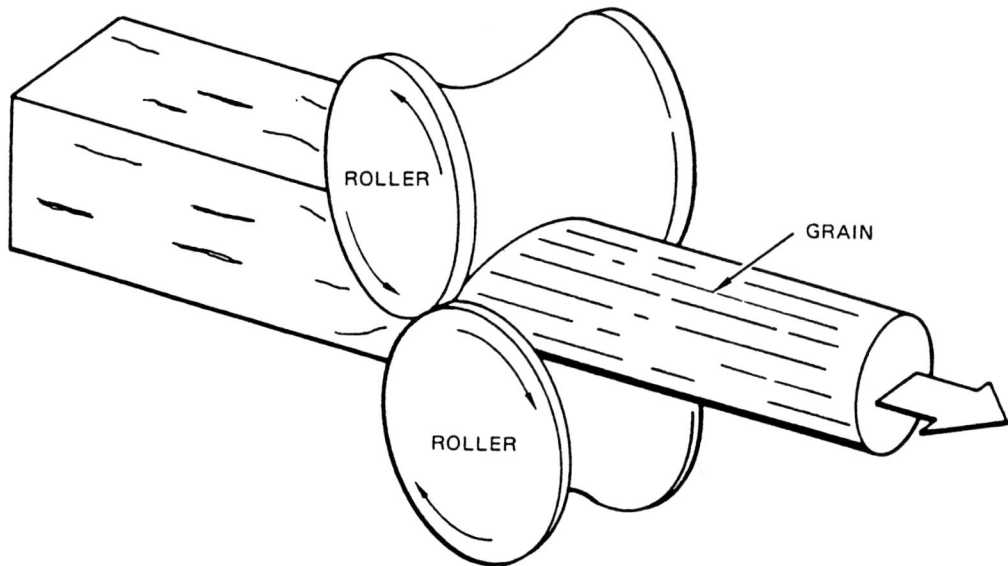

As the billet is rolled, it becomes longer and thinner. The crystal grain structure is broken up and re-forms into finer grain with a definite direction - in the direction of roll as shown above. The pockets of slag or nonmetallic inclusions are also stretched out in the direction of roll, becoming longer and thinner as does the grain structure.

Turn ahead to page 2-19.

From page 2-20   2-17

Up to this point, we have talked about discontinuities *within* the metal; nonmetallic inclusions, porosity, and pipe. These are known as SUBSURFACE discontinuities. Now, let us discuss discontinuities on the surface on the billet and how they react to the rolling process.

First let's discuss a common crack in the billet.

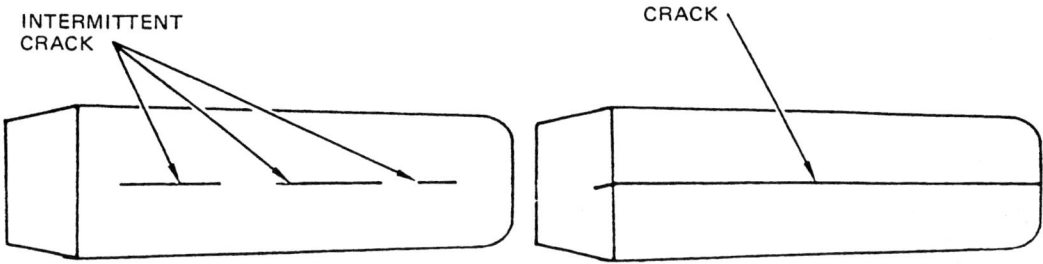

Like other discontinuities in the billet, the cracks get another name when the billet is rolled. As the billet is rolled and stretched out, so are the cracks. The cracks are then called SEAMS. SEAMS *are always open to the surface.*

Do you think the seams (cracks) are discontinuities?

Yes .... **Page 2-18**
No .... **Page 2-22**

Of course! A seam is an INTERRUPTION IN THE NORMAL PHYSICAL STRUCTURE OF THE BILLET; therefore, it is a discontinuity.

Seams are caused by surface cracks or irregularities on the ingot after it has been cast. Seams may also be caused by folding of the metal due to improper rolling or by a defect in the roller. As these discontinuities are rolled, they stretch out like taffy. As a result, the SEAM is lengthened in the direction of roll in the same way that nonmetallic inclusions are stretched out.

Turn ahead to page 2-21.

Right. A nonmetallic inclusion in bar stock extends in the direction of grain formation. As the billet is rolled smaller around and longer, the inclusion also becomes smaller around and longer. It is now called a STRINGER. A longitudinal cross section of the bar would show how the stringer is stretched out like this.

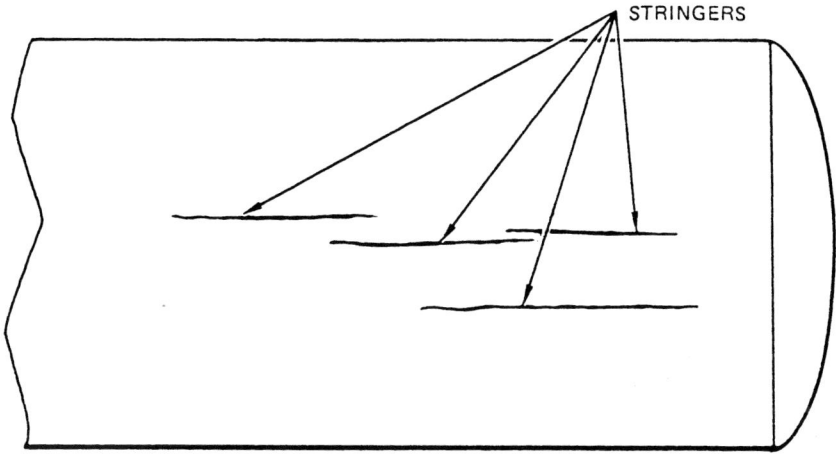

When a longitudinal cross section of the bar is cut, porosity and pipe would show up stretched out in the same manner as the stringers. However, they would appear as intermittent cracks or elongated cavities.

Turn to the next page.

Here you can see the shrinkage cavities or "pipe" in the billet. Notice how the pipe becomes smaller around and longer.

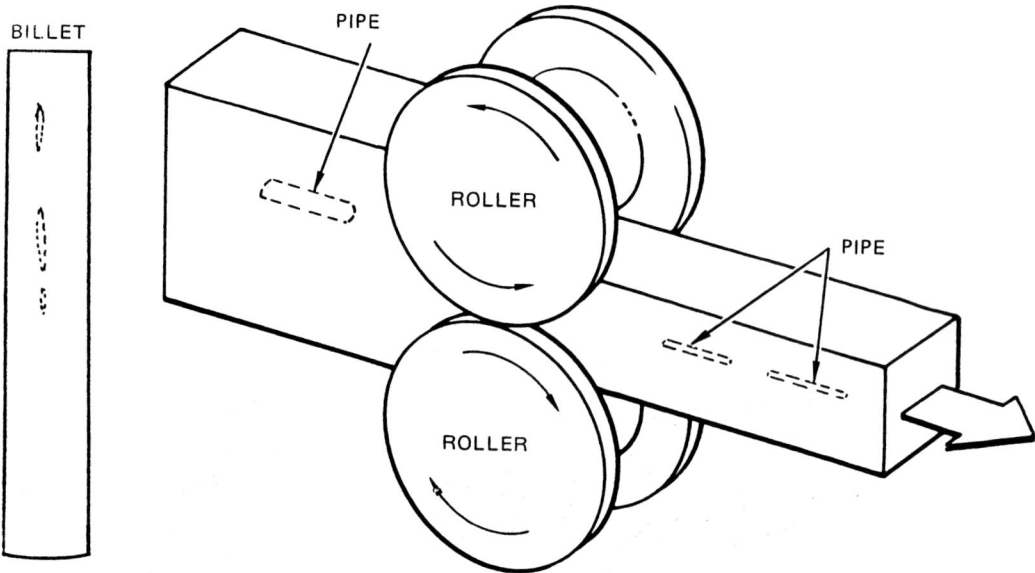

As the billet is rolled more and more, the pipe will continue to stretch out in the direction of roll. Pipe retains its name even in the elongated condition.

Turn back to page 2-17.

From page 2-18

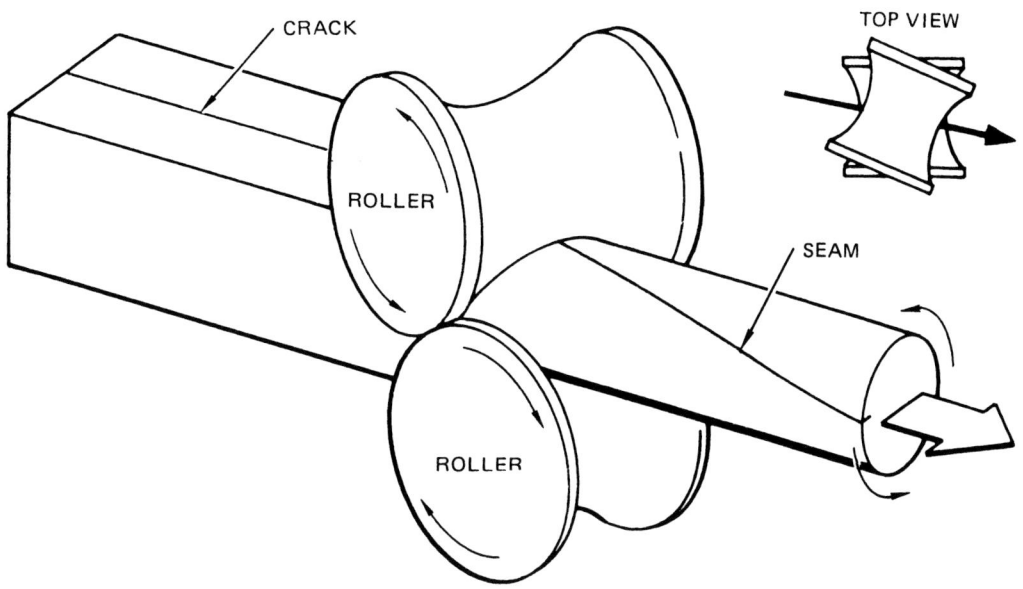

When the billet is rolled into bar stock, the seam may appear as a spiral in the bar. Such a spiral seam is caused by the angular mating of the rolls (see top view) which results in the material being turned as it is rolled. The example is exaggerated, of course.

Which of the following states the relationship between the direction of the seam and the direction of grain flow?

The seam forms at an angle to the grain flow .... **Page 2-23**
The seam forms in the direction of grain flow .... **Page 2-24**

From page 2-17

You don't think seams (cracks) in the billet are discontinuities. You must have forgotten the definition of a discontinuity, so let's take a quick look at it.

A DISCONTINUITY is defined as:

> A BREAK OR INTERRUPTION IN THE NORMAL PHYSICAL STRUCTURE OF AN ARTICLE.

A crack in a billet is not supposed to be there. Cracks are caused by shrinkage as the molten metal solidifies. So, as you see, cracks or seams are interruptions in the normal physical structure of the billet and are, therefore, DISCONTINUITIES.

Turn back to page 2-18.

You think the seam forms at an angle to the grain flow in that round bar. Actually, the turning tendency of the metal caused by the angular mating of the rolls will cause the grain to form in a spiral the same as the seam.

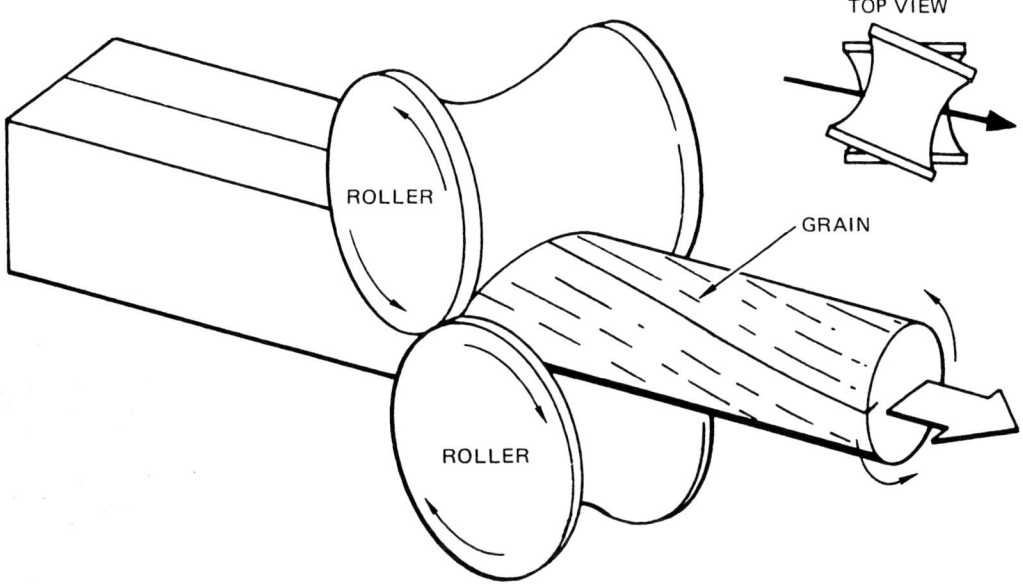

Remember that the illustration is exaggerated for the purpose of illustration. As the billet is rolled and re-rolled, the spiral effect becomes more pronounced. But the seam does form in the same direction as the grain.

Turn to the next page.

From page 2-21

Good for you. The seam forms in the direction of grain flow. The angular mating of the rolls causes the material to turn and the grain, as well as the seam, will be formed as a spiral.

Turn to the next page.

Now consider the cracked billet being rolled into bar stock.

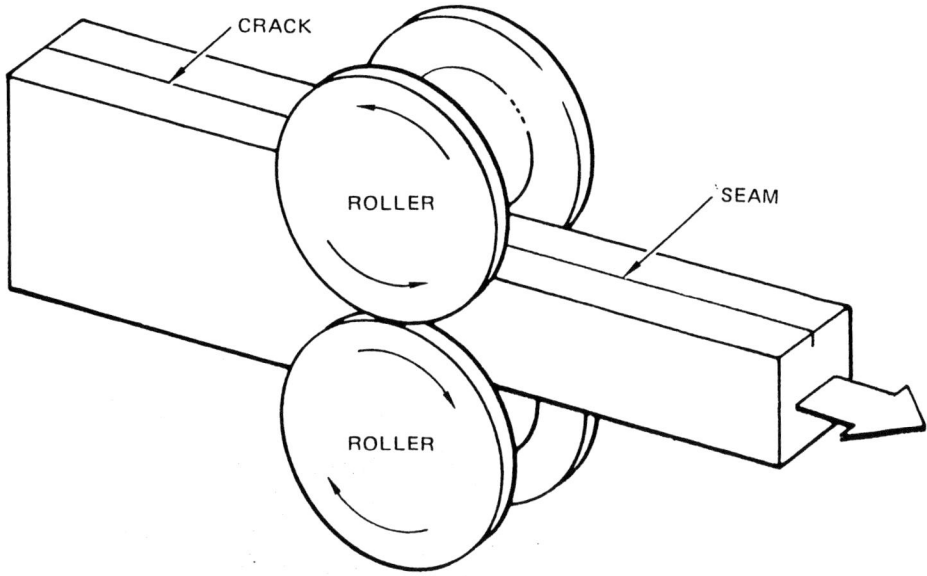

As you can see, a seam in the bar also stretches out in the direction of roll.

Considering the seam in the round rolled bar and the seam in the rectangular bar, which of the following correctly states the direction of grain flow in both cases?

In the direction of the seam in both cases .... Page 2-26
In the direction of the seam in the rectangular bar and
through the length of the round bar .... Page 2-28

Absolutely. The grain flows in the direction of the seam in both cases. The more the bar is rolled, the longer the seam will extend.

On a finished bar, a seam may appear as either a continuous straight line or a series of straight lines following each other.

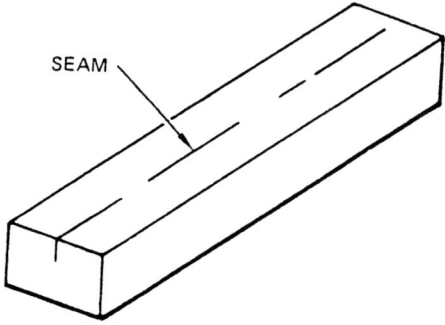

A seam in round bar stock or tubing also appears in a straight line or slight spiral which may be continuous or broken.

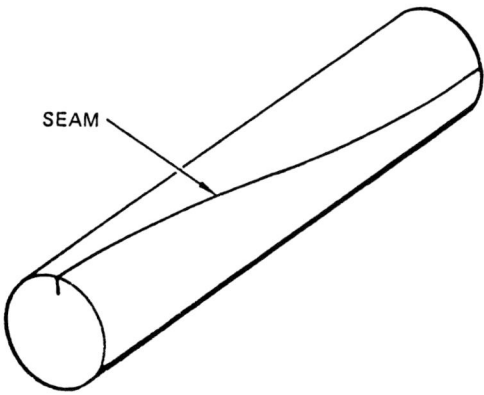

Turn to the next page.

From page 2-26

2-27

Any discontinuity will form in the direction of roll just as the grain does. The seam in the round bar had a slight spiral to it. The spiral is caused by the tendency of the rounded metal to spiral during rolling operations.

Which of the following types of discontinuities would you most likely expect to find in a SHEET OF METAL?

Seams	**Page 2-30**
Lamination	**Page 2-31**
Stringers	**Page 2-33**

From page 2-25

You think the grain flows in the direction of the seam in the rectangular bar and through the length of the round bar. You are half right. The grain does flow in the direction of the seam in the rectangular bar.

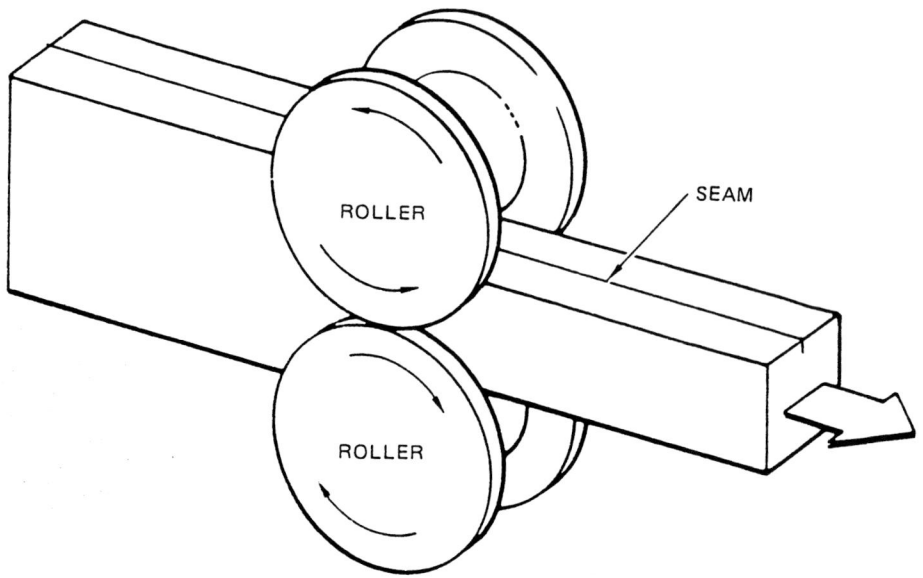

But when the billet is rolled into a round bar, both the seam and the grain form in a slight spiral. The angular mating of the rolls causes the material to have a turning tendency.

Turn to the next page.

From page 2-28

2-29

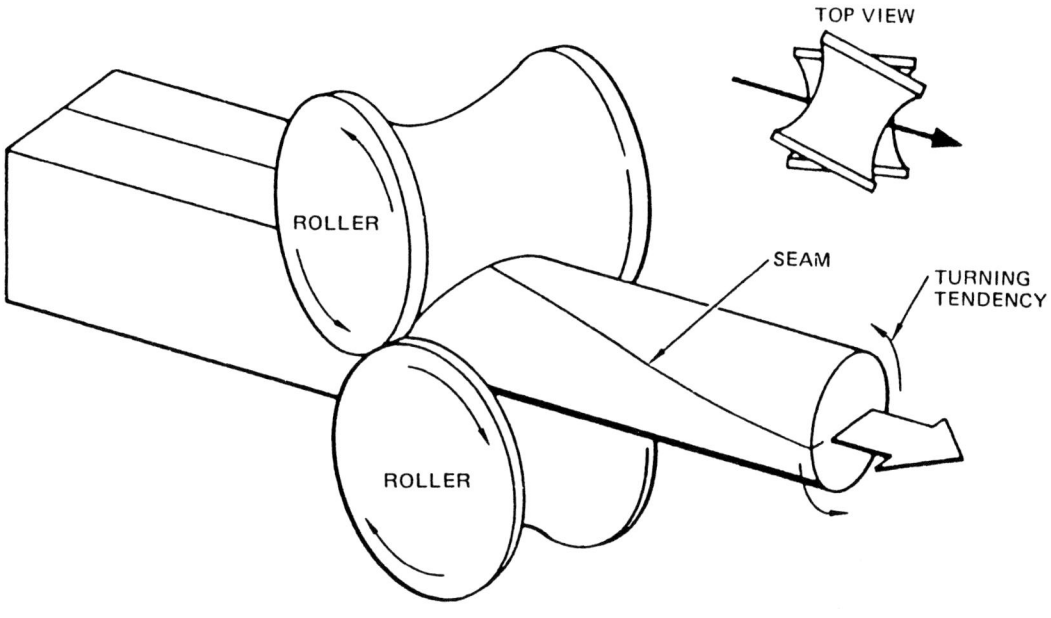

So you see, the grain flows in the direction of the seam in both cases.

Turn back to page 2-26.

From page 2-27

You would expect to find "seams" in a sheet of metal. Probably not. Seams are usually found in the surface of bar stock. See if this illustration won't give you a clue to the correct answer.

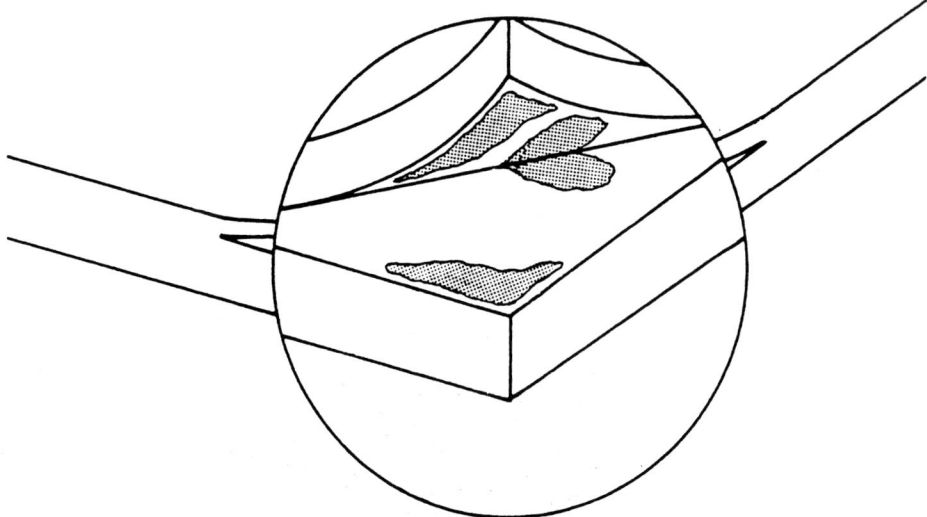

In a sheet of metal, you would expect to find nonmetallic inclusions or gas porosity that have been flattened by rollers and that have spread out in all directions, but mainly in the direction of roll.

Turn back to page 2-27 and select the correct answer.

Yes, of course. You would expect to find laminations in a sheet of metal. A lamination is simply a nonmetallic inclusion that has been rolled or squashed and spreads out in many directions, although mainly in the direction of roll. Of course gas-bubble porosity will also cause laminations when rolled into sheet stock.

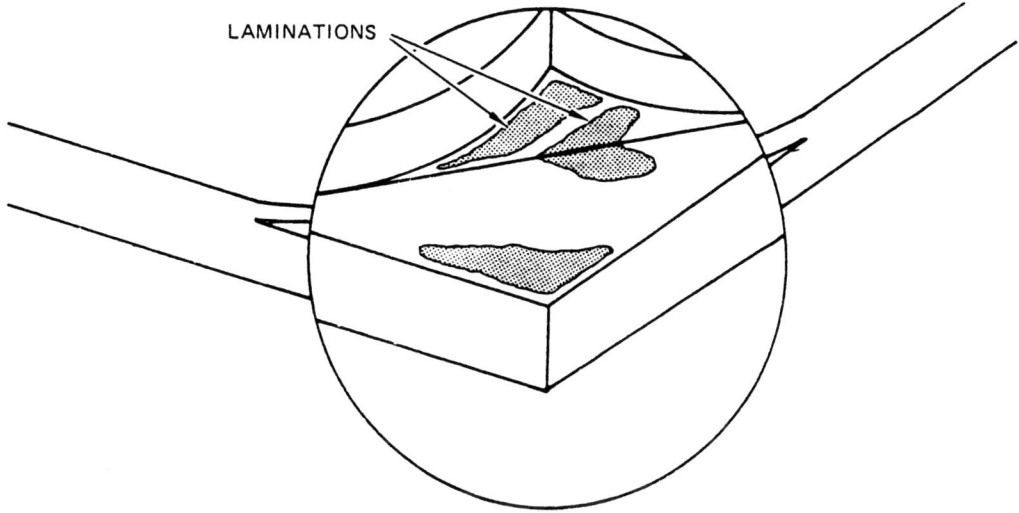

Turn to the next page.

From page 2-31

Seams and stringers are discontinuities that have been stretched out by rolling and are found in bar stock.

Before a billet is rolled into *bar stock*, it may contain subsurface foreign inclusions. When the billet is rolled into bar stock, these inclusions get another name.

What are the inclusions then called?

Seams .... Page 2-34
Nonmetallic inclusions .... Page 2-35
Stringers .... Page 2-37

From page 2-27

You would expect to find stringers in a sheet of metal. No, stringers would probably not be found in a sheet of metal, and here is why.

Stringers are formed in the rolling process from nonmetallic inclusions in the billet. The inclusions are pockets of foreign materials in the billet. When the billet is rolled, the metal becomes smaller around and longer. The nonmetallic inclusions are also squeezed and stretched out like taffy.

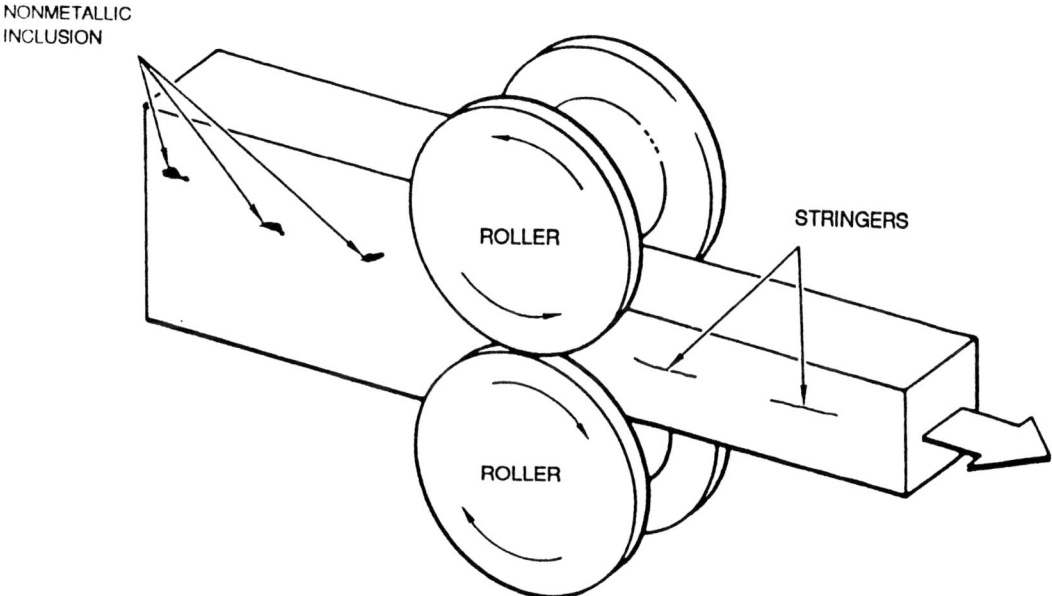

Here you can see how stringers are formed from nonmetallic inclusions. The more times the bar is rolled, the longer and thinner the stringers become. In the processing of sheet materials, the rolling operation will tend to flatten rather than just elongate the inclusion. The result is a lamination rather than a stringer.

Turn back to page 2-27 and try again.

From page 2-32

You think nonmetallic inclusions in the billet are called seams after the billet is rolled into bar stock. No, seams are caused by cracks in the surface of the billet.

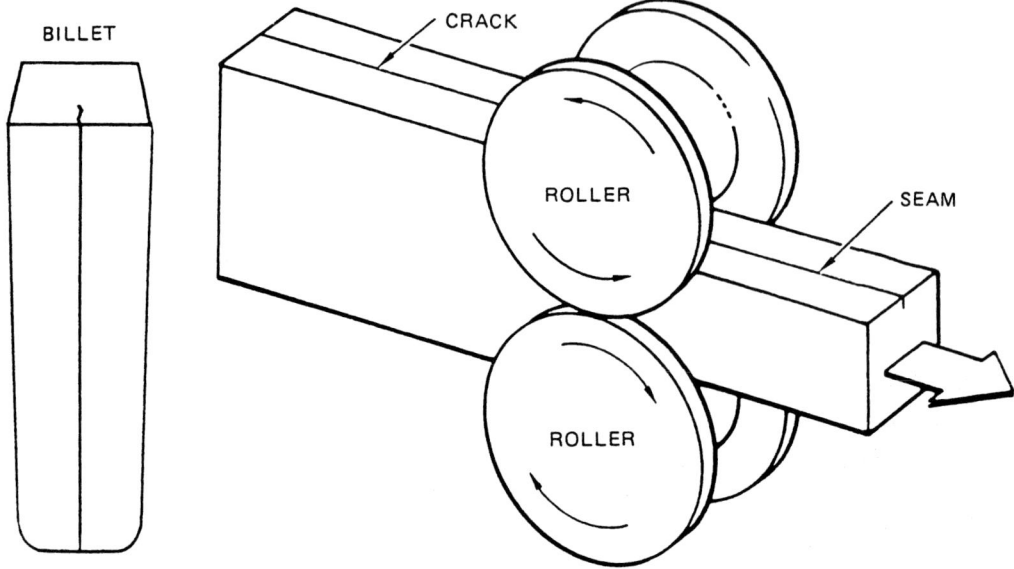

Another important fact here is that seams are always open to the surface. Nonmetallic inclusions are inside the billet; they are subsurface.

Turn back to page 2-32 and re-read the last paragraph.

You think nonmetallic inclusions in the billet have the same name after the billet is rolled. Remember now that a nonmetallic inclusion in the billet is an irregularly-shaped pocket of impurities. After the billet is rolled, the nonmetallic inclusion is stretched out and becomes longer and thinner and forms in the direction of roll. In plate material, the inclusion is flattened out in all directions, but mainly in the direction of roll. It is then called a LAMINATION.

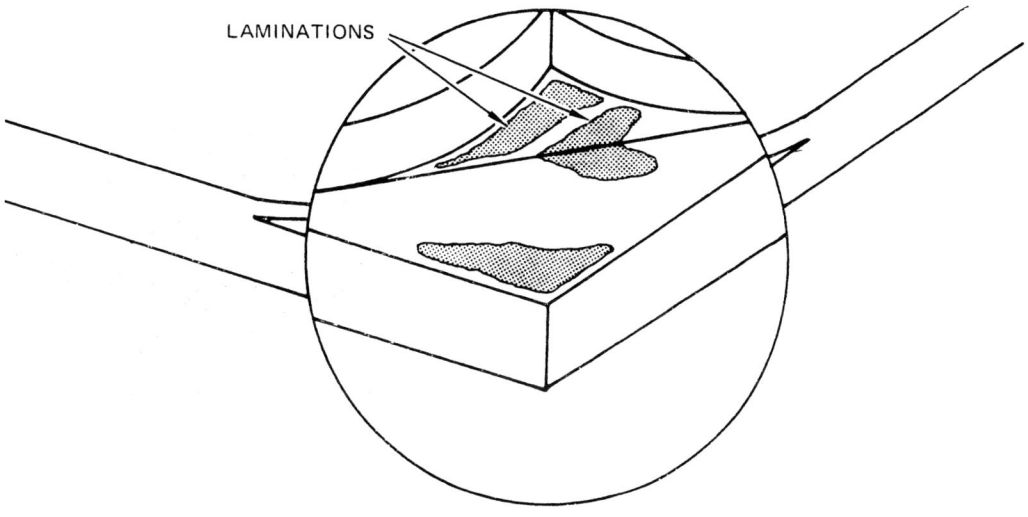

Turn to next page.

From page 2-35

In bar stock, the nonmetallic inclusion is made longer and thinner.

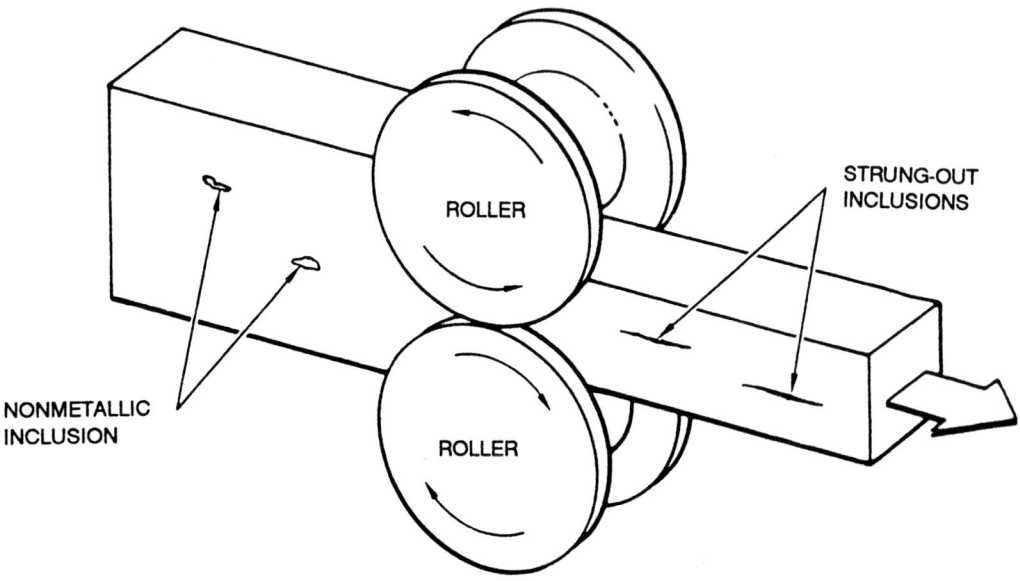

After being rolled, those nonmetallic inclusions are strung out.

Turn back to page 2-32 and select the correct answer.

From page 2-32

You are right. The nonmetallic inclusion is called a stringer. Of course, the nonmetallic material is still there, but it is stretched out. In a cross section of bar stock, the stringer would look like this.

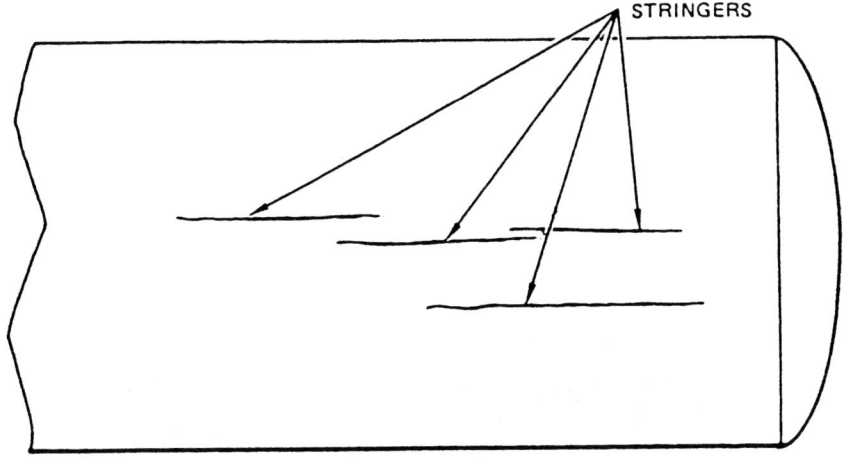

Turn to the next page for a short review.

CHAPTER REVIEW

____ 1. It is important to remember that the cropped ingot or billet has a crystal grain structure. In the billet, the grain:

    A. has no direction.
    B. has a direction.
    C. is independent of direction of processing.
    D. structure is destroyed.

____ 2. Working or rolling of the billet breaks the grain structure down. The grain becomes finer and begins to have:

    A. laminations.
    B. discontinuities.
    C. hot tops.
    D. direction.

____ 3. The grain forms in the direction in which the metal is:

    A. pointing.
    B. rolled.
    C. heated.
    D. cropped.

From page 2-38

4. Before the billet is rolled, it may have discontinuities in the metal such as nonmetallic inclusions, pipe, and:

   A. gas porosity.
   B. metallic porosity.
   C. metallic hot tops.
   D. laminations.

5. When the slab is rolled into sheet or plate stock, nonmetallic inclusions are flattened but:

   A. spread mainly in the direction of heating.
   B. shrink in the direction of rolling.
   C. spread in the direction of rolling.
   D. shrink in the direction of cropping.

6. Inclusions in plates are sandwiched in the flattened-out metal and are called:

   A. stringers.
   B. inclusions.
   C. hot spots.
   D. laminations.

7. When the billet is made into bar stock, the nonmetallic metal inclusions are stretched out like taffy and are called:

   A. stringers.
   B. laminations.
   C. candy lines.
   D. bar lines.

___ 8. Stringers (stretched-out inclusions) form _____ _____ direction just as the grains do.

    A. across the rolling
    B. with the rolling
    C. independent of the rolling
    D. with the heating

___ 9. Pipe and gas-bubble porosity also form laminations when the billet is rolled into:

    A. stringers.
    B. laminations.
    C. bar.
    D. sheet or plate stock.

___ 10. When the billet is rolled into bar stock, the shrinkage cavities become smaller around and longer and are called:

    A. pipe.
    B. porosity.
    C. laminations.
    D. blooms.

11. Since nonmetallic inclusions, pipe, and porosity are found in the metal, they are:

    A. surface connected.
    B. considered unimportant.
    C. subsurface discontinuities.
    D. the only important problems in steel making.

12. Surface discontinuities (open to the surface) originate from surface irregularities or cracks in the:

    A. blast furnace.
    B. billet.
    C. pipe.
    D. finished product.

13. When the billet is rolled and stretched out, any cracks in the billet are also stretched out. The cracks are then called:

    A. billet lines.
    B. blooms.
    C. cracks.
    D. seams.

14. Seams are:

   A. not open to the surface.
   B. often open to the surface.
   C. always open to the surface.
   D. subsurface discontinuities.

15. Seams form:

   A. with the rolling direction.
   B. across the rolling direction.
   C. randomly on the part's surface.
   D. under the part's surface.

Turn ahead to page 2-44 for answers to the review questions.

From page 2-42

Turn to the next page.

From page 2-43   2-44

ANSWERS TO REVIEW QUESTIONS
FOR CHAPTER 2

Question & Answer		Reference Page(s)
1.	A	2-1
2.	D	2-2
3.	B	2-5
4.	A	2-1
5.	C	2-8
6.	D	2-6
7.	A	2-33
8.	B	2-33
9.	D	2-12
10.	A	2-20
11.	C	2-17
12.	B	2-17
13.	D	2-17
14.	C	2-17
15.	A	2-18

Turn to the next page and continue.

CHAPTER 3

FORGING DISCONTINUITIES

In the previous section we discussed working of the slab or billet and the effects the forming process has on grain structure and discontinuities. In this section we are going to discuss another metal-forming process called "forging."

Forging is the working of metal into a desired shape by hammering or pressing the metal while it is forced between rigid objects designed to provide the final shape desired. These rigid objects can be open or closed dies, and rollers of a variety of sizes and shapes. Though forging has long been considered a process of shaping hot, soft metal, current forging operations allow the shaping of metal in a cold state. For instance, the forging process known as "coining" shapes a workpiece by totally confining it in dies, pressing or loading the dies, and maintaining this pressure for a period of time until the desired shape is attained. This process, as well as many of the forging processes, can be automated.

Since some discontinuities are caused by the forging process itself, let's take a brief look at the closed-die, hot-forging process. Many of the discontinuities associated with this process are similar to other forging processes.

Turn to the next page.

The forging process starts with forging stock. These drawings are sliced down the middle so you will be able to see what goes on inside.

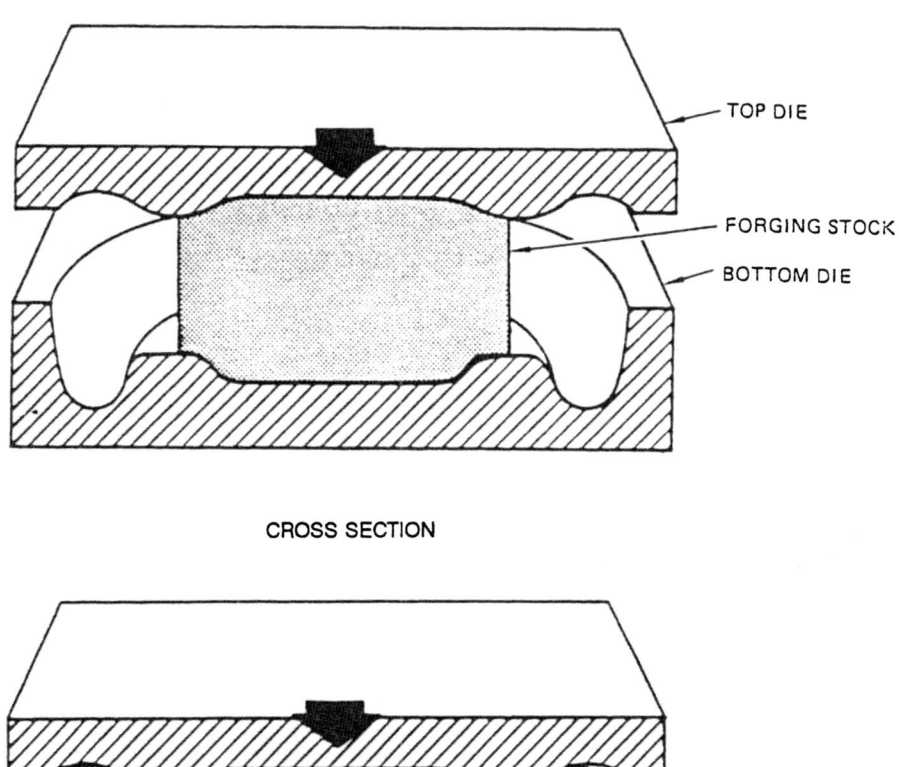

CROSS SECTION

The forging operation uses two dies. The billet is gradually heated to forging temperature (heat-soaked) and then placed between the two dies. The forging press or hammer then squeezes or pounds the hot metal between the two dies. Forging the metal breaks down the crystal grain structure to form finer grain just as in rolling operations.

Turn to the next page.

From page 3-2   3-3

The forging process takes several steps.

The first step is to squeeze or hammer the billet into a roughing die. This gives the metal a rough shape of the desired article. Here you can see the hot metal is rough shaped from being squeezed or pounded between the discs.

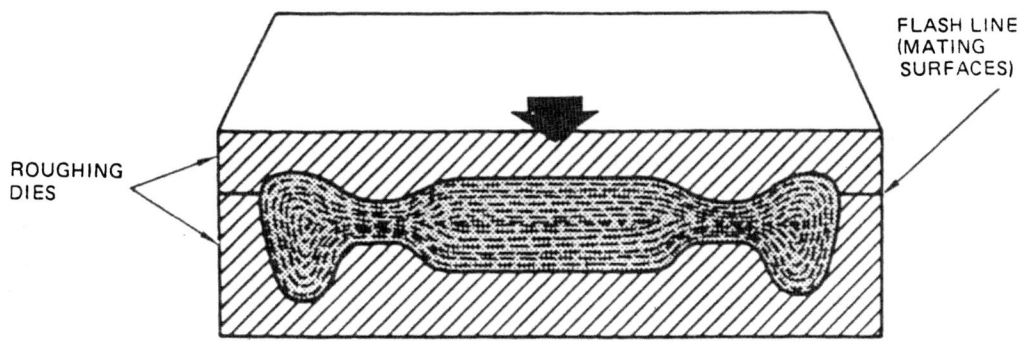

Working the metal by hammering it into the two dies breaks the crystal grain structure down into finer grain. The grain follows the form of the dies in the same manner as rolling operations where the grain follows the direction of roll.

After the metal is rough shaped, it is placed in a finishing die and is hammered into a more refined version of the same shape.

What do you think happens to the grain?

The grain will follow the more refined shape . . . . Page 3-5
The grain will remain the same . . . . Page 3-7

From page 3-8   3-4

A FORGING LAP IS A DISCONTINUITY CAUSED BY FOLDING OF METAL IN A THIN PLATE ON THE SURFACE OF THE FORGING. One type of forging lap is caused when the mating surfaces of the two forging dies do not match.

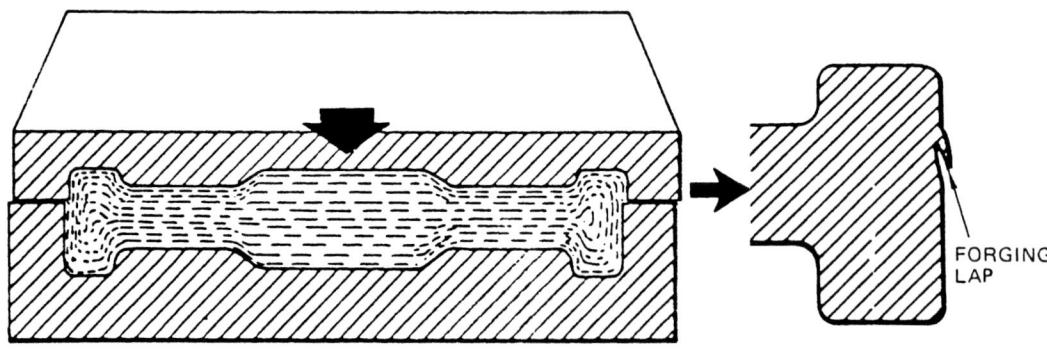

As the forging stock is squeezed down between the two dies, some of the metal will be forced out between the mating surface. As the metal is further squeezed down, the metal will be sheared and folded up into a lap on the surface of the forging.

Since the forging lap is caused by a faulty die, do you think that it is *always* open to the surface?

No .... **Page 3-9**
Yes .... **Page 3-10**

That's right. The grain will follow the more refined shape. The more the metal is worked, the finer the grain will become. And the grain will follow the shape into which the object is being made.

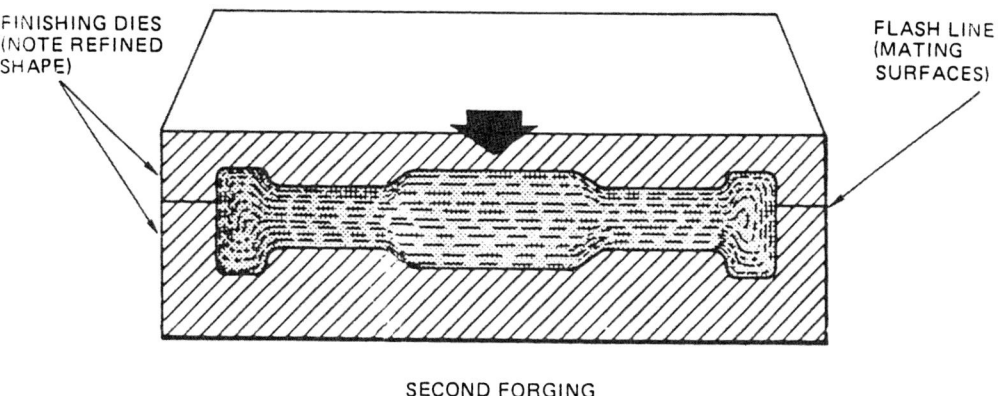

SECOND FORGING

In summary then, the forging metal is placed in the lower die. The upper die is then brought down in repeated blows - each blow making a change in the shape of the forging. The impact and pressure cause the metal to fill the die.

Grain flow of the metal undergoes a similar change into a repetition of the shape of the forging contours. The greatest value of this characteristic is that the grain flow lines remain unbroken, and the forging is unified into a continuous structure, strong and tough throughout.

What is the main advantage of the grain flow lines in a forging?

The grain flow lines make a piece of metal easier to forge .... **Page 3-6**
The grain flow lines provide strength .... **Page 3-8**

From page 3-5

You believe that the main advantage of grain flow lines is to make a piece of metal easier to forge. That's not the answer we were looking for.

As metal is forged, the grain structure conforms to the shape of the object being formed. The grain flow lines remain unbroken as the metal is forged. In this manner, the metal forms a unified, continuous structure that is consistently strong throughout. Therefore, we would say that the main advantage of the grain flow lines is that they . . .

. . . provide strength . . . . Page 3-8

From page 3-3

You believe the grain of a metal part forged in a finishing die retains the same flow as it had during the rough forging stage. That's not exactly so.

Working metal by hammering it between two dies breaks the crystal grain structure down into finer grain, whether the dies are roughing dies or finishing dies. And, as the grain is broken down, it *follows the form of the finishing die*.

Turn back to page 3-5.

Absolutely. The grain flow lines provide strength to the forging. Since the grain flow lines remain unbroken, the forging is unified into a continuous structure, which is strong and tough.

Here is a photograph of the grain flow lines in a cross section of an aluminum forging.

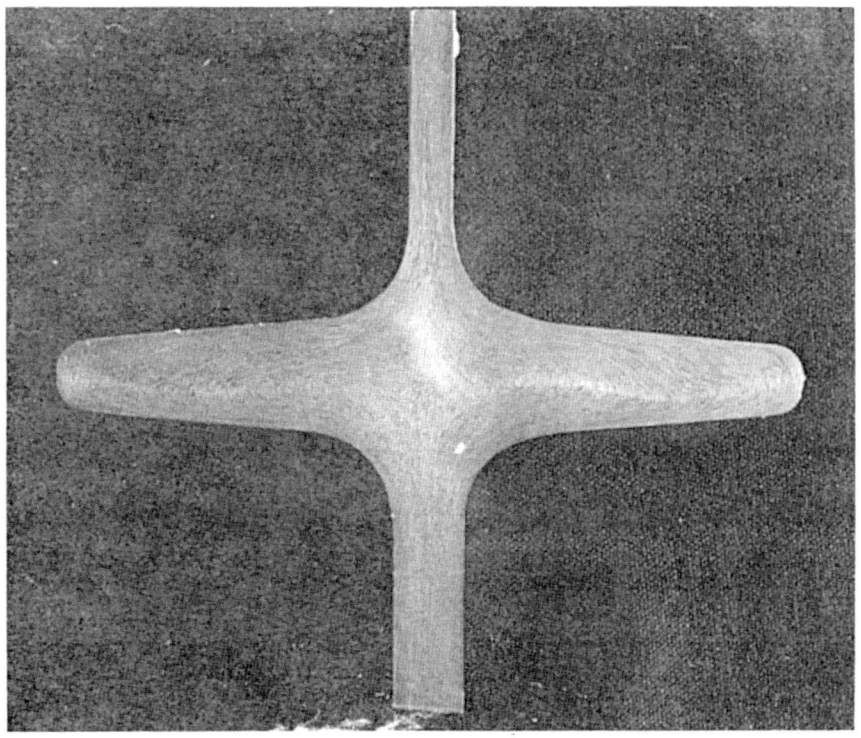

There are two primary types of discontinuities formed as a result of forging:

- FORGING LAPS and
- FORGING BURSTS OR CRACKS

Turn back to page 3-4.

From page 3-4

Actually, the answer is "yes." Let's take another look at the definition of a forging lap.

> A FORGING LAP IS A DISCONTINUITY CAUSED BY FOLDING OF METAL IN A THIN PLATE ON THE *SURFACE* OF THE FORGING.

Forging laps are always open to the surface.

Turn to the next page.

From page 3-4   3-10

Yes, of course. Forging laps are always open to the surface, and they may be found at the point where the two dies come together. Let's take another look at the definition of a lap.

> A FORGING LAP IS A DISCONTINUITY CAUSED BY FOLDING OF METAL IN A THIN PLATE ON THE SURFACE OF THE FORGING.

Here is our forging taken out of the die. It has been cut in half so you can see the inside of the forging.

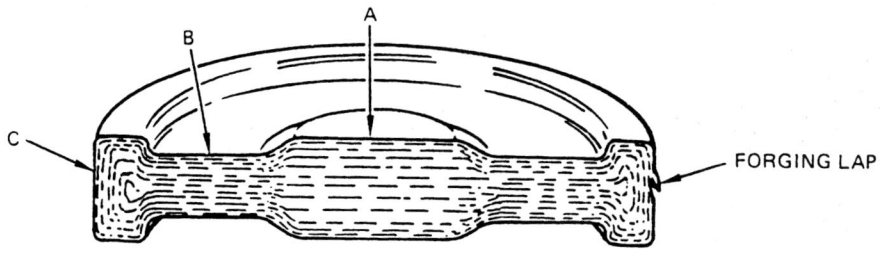

Since the forging lap shown above was made at the mating surface of the two forging dies, where else on the forging would there be another forging lap of this type (point A, B, or C)?

A .... Page 3-12
B .... Page 3-14
C .... Page 3-16

From page 3-16

You selected A, and that is incorrect. You see, there is no abrupt change in grain direction at point A.

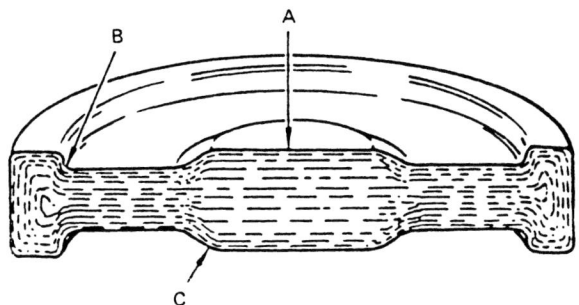

At point C, the grain does have some change in direction, but the change is not very abrupt. Grain direction changes are most abrupt where there is a sharp change in the profile of the forged object.

Turn ahead to page 3-16 and try again.

From page 3-10 3-12

You think that another forging lap would be located at point A. Evidently we haven't made the point clear. Let's take another look at the problem. First, here is the definition to consider.

> A FORGING LAP IS A DISCONTINUITY CAUSED BY FOLDING OF METAL IN A THIN PLATE ON THE SURFACE OF THE FORGING.

Now, here is our forging in the dies again.

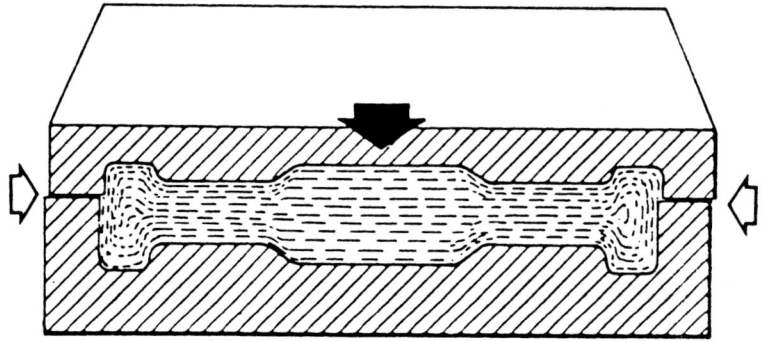

There are only two places where the dies come together - one on the right and one on the left.

Turn back to page 3-10 and select another alternative.

From page 3-16   3-13

Good for you. The most abrupt change in grain direction was shown at point B.

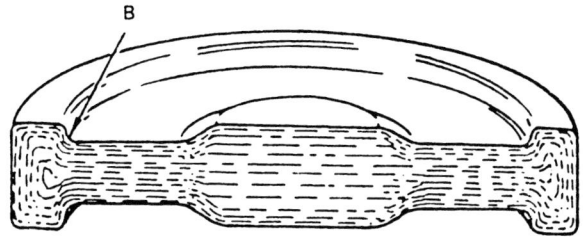

A forging lap is most likely to occur where the grain makes the most abrupt change in direction. At the points where the grain makes the most abrupt change in direction, the grains are also close together.

So far, we have shown only a cutaway view of this forged part. We have been forging a gear blank. Here is what the complete part looks like.

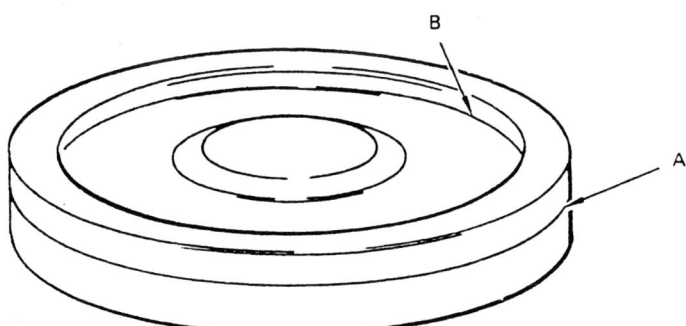

At which point in the picture above would a forging lap most likely be caused by the mating surface of the two forging dies?

A . . . . Page 3-17
B . . . . Page 3-18

From page 3-10 3-14

You selected B, and that's not quite it. Let's take another look at this thing. First, here is the definition of a forging lap.

> A FORGING LAP IS A DISCONTINUITY CAUSED BY FOLDING OF METAL IN A THIN PLATE ON THE SURFACE OF THE FORGING.

Now, let's take another look at our forging in the dies.

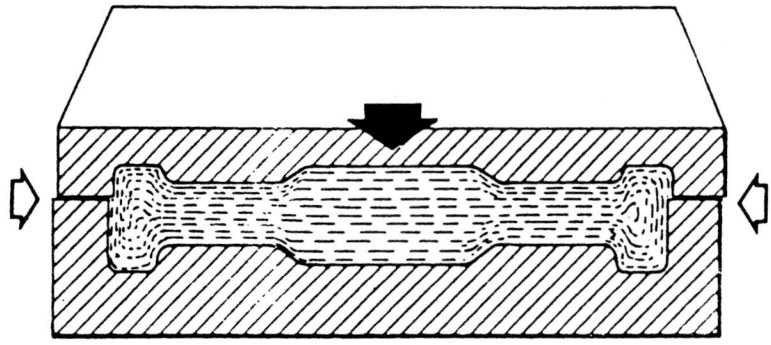

There are only two places where the dies come together as mating surfaces to cause a lap - one is on the right and the other on the left.

Turn back to page 3-10 and try another answer.

From page 3-16

You selected C. It is evident that you are getting the idea. There is some change in grain direction at point C, but it is not very abrupt. Therefore, it would be very unlikely to cause a forging lap at that point in the forging.

Turn ahead to page 3-16 and try once again.

From page 3-10   3-16

Right. Another forging lap could be found at point C.

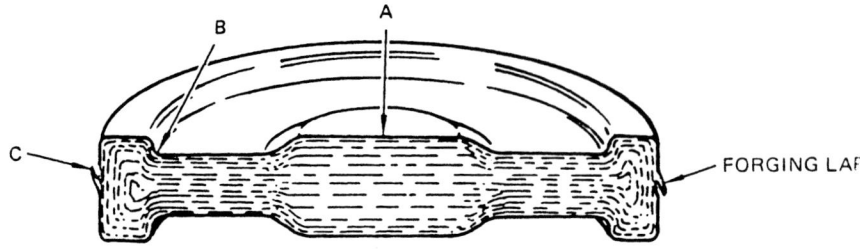

Point C is the other point where the two forging dies come together. The forging lap will always be open to the surface.

Another type of forging lap is formed in a forging at points where there is an abrupt change in grain direction. Here is our forging showing grain structure.

At which of the following points on the forging is the most abrupt change in grain direction shown?

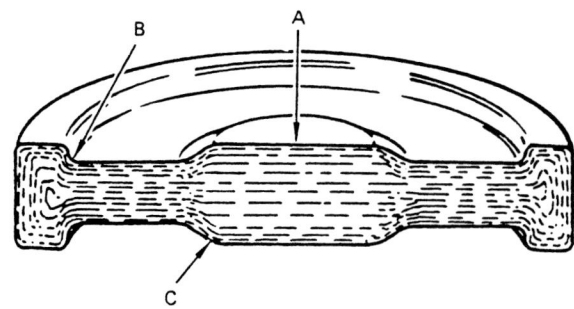

A . . . . Page 3-11
B . . . . Page 3-13
C . . . . Page 3-15

From page 3-13   3-17

Right. Point A is the place where a forging lap may be caused by the mating surfaces of the two forging dies.

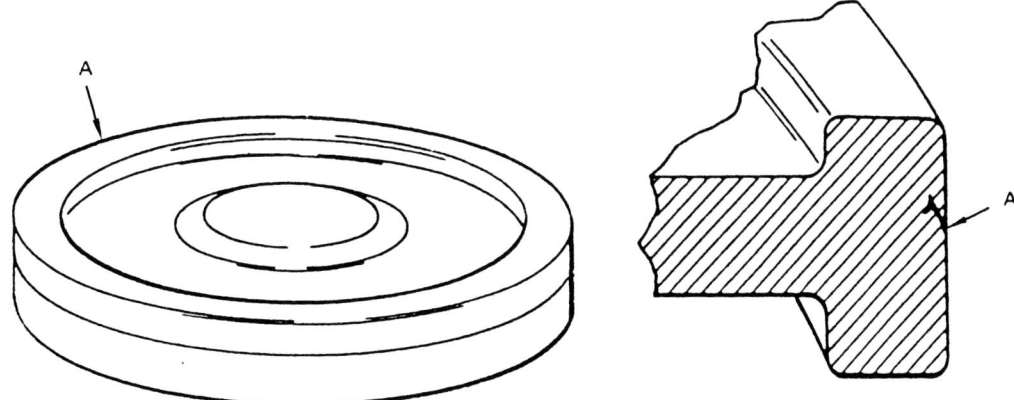

Now, before we go on, select the statement below which is the definition of a forging lap.

A FORGING LAP IS A DISCONTINUITY CAUSED BY FOLDING OF METAL IN A THIN PLATE WITHIN THE METAL FORGING .... **Page 3-19**

A FORGING LAP IS A DISCONTINUITY CAUSED BY FOLDING OF METAL IN A THIN PLATE ON THE SURFACE OF THE FORGING .... **Page 3-20**

From page 3-13   3-18

You are correct inasmuch as a forging lap can occur at point B due to an abrupt change in grain direction.

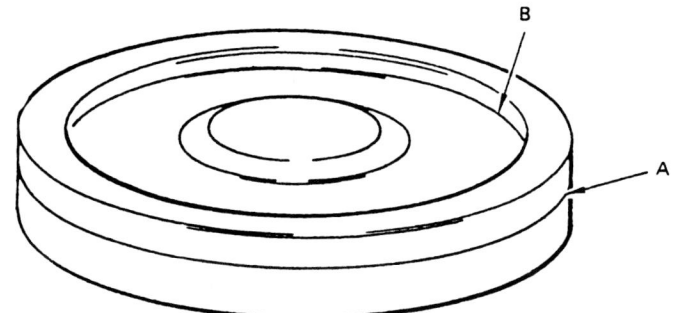

At point B, only the top surface die is involved. So, a forging lap caused by mating surfaces is not possible. In the case of our forging, there is only one designated point where the dies come together, and that is point A.

Turn back to page 3-17.

You selected A FORGING LAP IS A DISCONTINUITY CAUSED BY FOLDING OF METAL IN A THIN PLATE WITHIN THE METAL FORGING. Let's take a look at a forging lap.

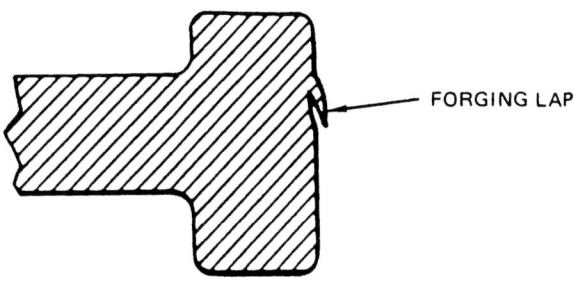

Actually, you can see that the forging lap is open to the surface. In fact, all forging laps are open to the surface. Here is the correct definition of a forging lap.

> A FORGING LAP IS A DISCONTINUITY CAUSED BY FOLDING OF METAL IN A THIN PLATE ON THE SURFACE OF THE FORGING.

Turn to the next page.

From page 3-17   3-20

You are absolutely right. The definition of a forging lap is:

A DISCONTINUITY CAUSED BY FOLDING OF METAL IN
A THIN PLATE ON THE SURFACE OF THE FORGING.

Forging laps can also be caused by poor die design. As the metal is pressed into the cavity in this die, the metal is forced up at the bottom of the die and tends to fold over on itself, forming the forging lap shown on the right.

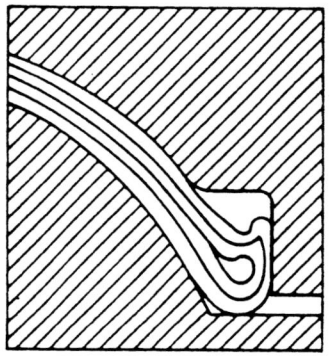 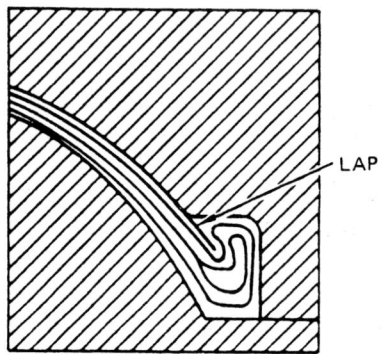

In a properly engineered die, there will be no forging lap.

Turn to the next page.

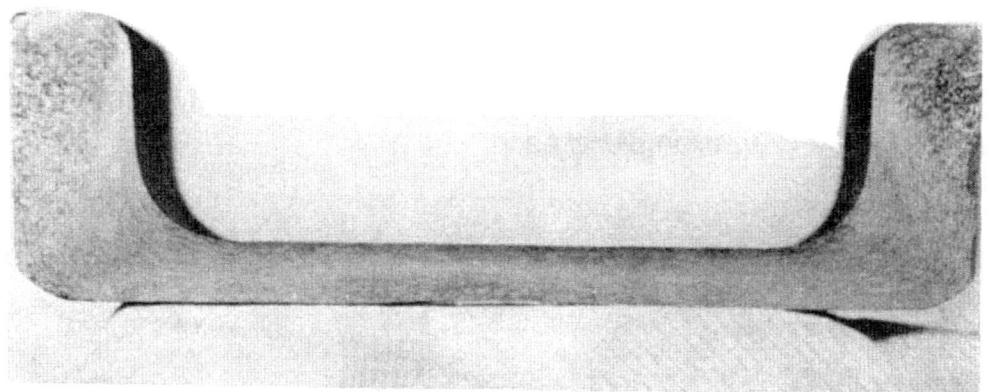

This picture shows the grain flow in a titanium forging that does not have a forging lap. Remember, a forging lap is always open to the surface.

Now let's discuss FORGING BURSTS, which may be either open to the surface or subsurface.

Turn to the next page.

From page 3-21   3-22

A FORGING BURST IS A RUPTURE CAUSED BY FORGING AT IMPROPER TEMPERATURES. Forging a metal at too low a temperature may cause bursts. These bursts may be either internal or they may occur at the surface. Here is an example of each.

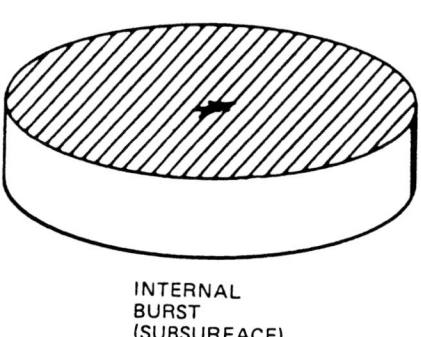

INTERNAL
BURST
(SUBSURFACE)

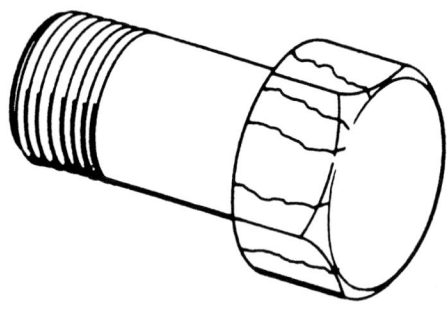

EXTERNAL
BURST OR CRACK
(OPEN TO THE SURFACE)

Improper temperatures caused these parts to break as the material was being shaped by forging. The metal was not hot enough and did not want to flow with the forging. When squeezed by the heavy forging press, the metal in the center simply ruptured.

Which of the following conditions would likely cause a burst?

Overheating metal . . . . **Page 3-24**
Underheating metal . . . . **Page 3-26**

"Yes" was your selection. We haven't made the point clear. Let's take another look.

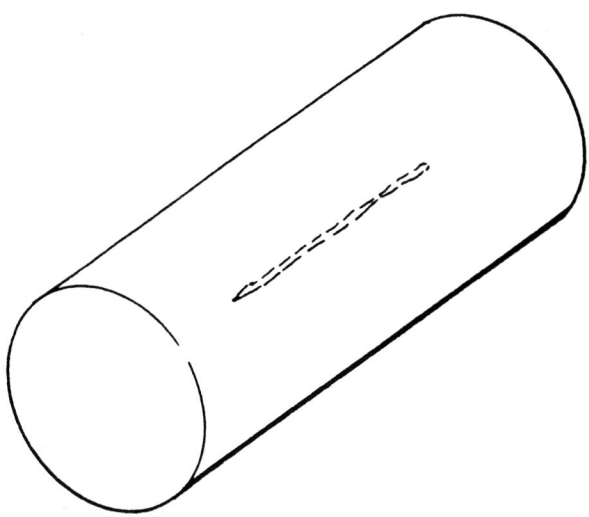

Here is the forging before we cut the slice out of the center. If we could see inside of the forging, the burst would look like the above.

We would not expect to find the internal burst open to the surface like a forging lap, because only the center was not heated sufficiently. The rest of the material was heated properly. Because the center was not hot enough, the metal did not want to flow with the forging. It simply ruptured or cracked in the center where it was not quite hot enough. So you wouldn't expect to find an internal burst like this one open to the surface like a forging lap.

Turn ahead to page 3-28.

From page 3-22

You selected overheating metal, and that's not quite right. If the metal is overheated, it tends to be softer. If the metal is soft, it can more easily be worked into the shape of the dies.

If the metal is underheated, it is not as soft. The metal does not flow easily into the shape of the dies. As a result, the metal cracks as it is worked.

Turn ahead to page 3-26.

Of course not! You wouldn't expect all forging bursts or cracks to appear only in the centers of forgings. There are both internal and external bursts. Here are two examples.

The photo on the left shows external bursts or cracks on the surface of the forging. On the right is an internal burst found in a forged bar of titanium.

Turn ahead to page 3-29.

From page 3-22   3-26

You bet. Underheating of the metal would most likely cause a burst. Let's take a look at an internal burst.

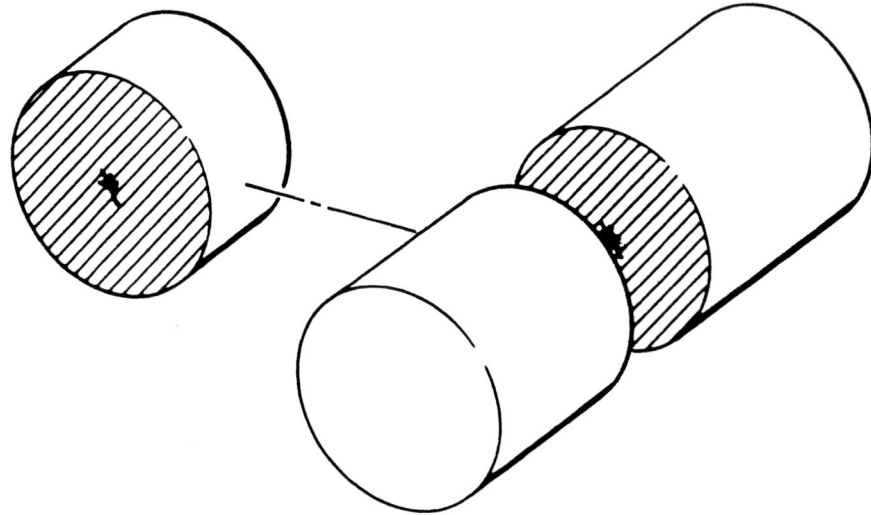

We have sliced a piece of metal out of the center of the forging so that you can see the burst in the center.

The center of this large piece of metal was not heat-soaked to the proper forging temperature before the forging operation began. Because the center was not hot enough, the metal did not want to flow with the forging. It simply ruptured or cracked in the center when squeezed by the heavy press.

Would you expect to find the above internal forging burst open to the surface like a forging lap?

Yes .... **Page 3-23**
No .... **Page 3-28**

From page 3-28

"Would you expect to find all forging bursts or cracks only in the centers of the forgings?" You answered "Yes," which is incorrect.

A forging burst or crack might occur in the center of the metal, or it might have a surface opening.

Below is the picture of the bolt with the surface cracks that we discussed earlier. Remember?

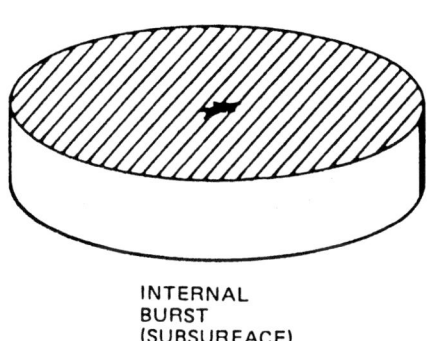

INTERNAL
BURST
(SUBSURFACE)

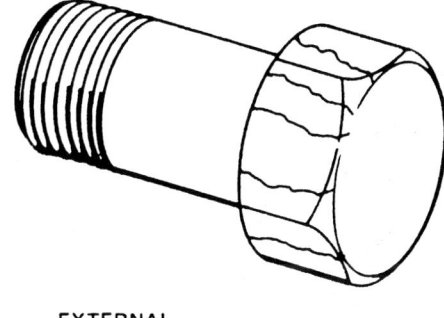

EXTERNAL
BURST OR CRACK
(OPEN TO THE SURFACE)

Turn back to page 3-25 and continue.

From page 3-26   3-28

"No" is the correct answer. An internal forging burst would not be open to the surface like a forging lap. To see an internal burst, you would have to machine away some of the metal. Here is an actual picture of that internal burst.

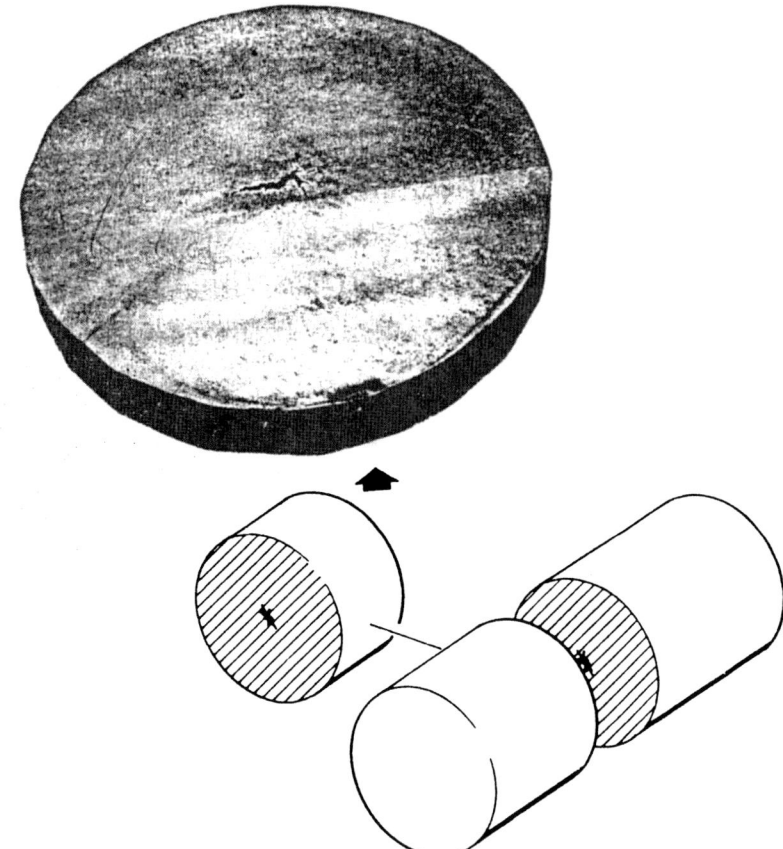

As you can see, the burst is very ragged and is close to the center of the metal.

Would you expect to find all forging bursts or cracks only in the center of forgings?

No .... **Page 3-25**
Yes .... **Page 3-27**

From page 3-25

In summary, forging discontinuities can be caused when the metal being forged has not been heated throughout to the proper temperature. That portion which has not been sufficiently heated will not flow properly and may rupture. These discontinuities can be subsurface or open to the surface. These discontinuities are called forging bursts or forging cracks.

To review briefly, grain flow of a forging follows the shape of the forging contours. This characteristic is important because the grain flow lines remain unbroken and because the forging is unified into a continuous structure, which is strong and tough.

Turn to the next page.

Here are some examples.

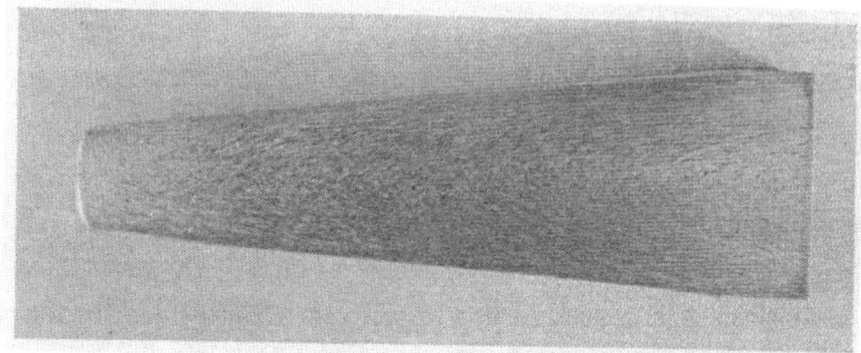

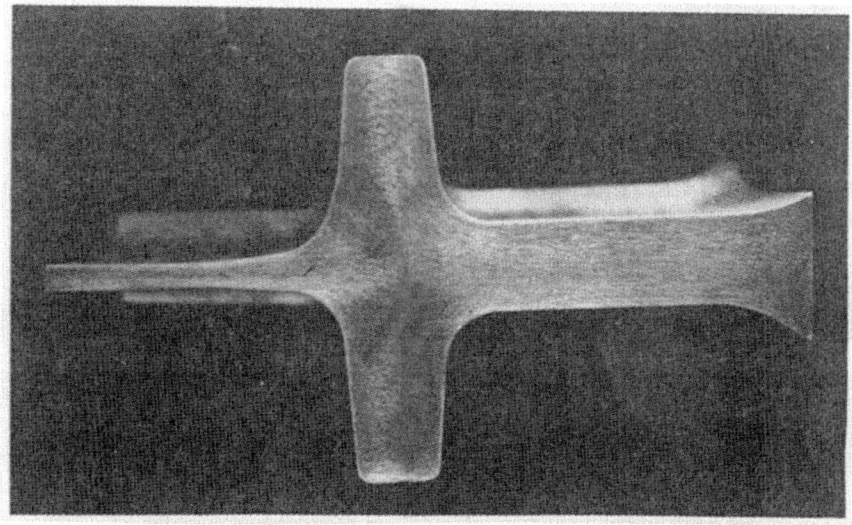

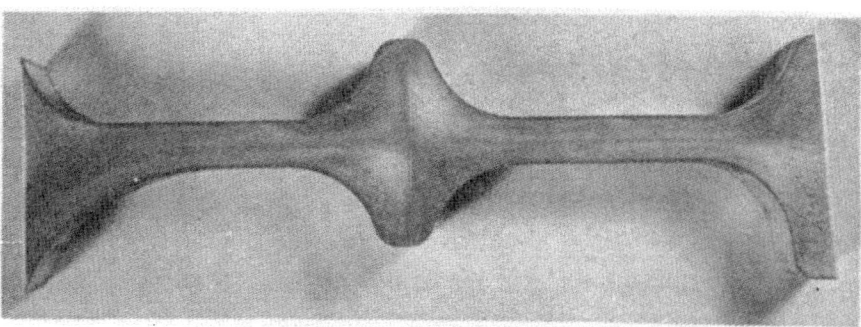

Turn to the next page.

High-strength precision forgings of intricate design and small sizes are now available. Here are a few examples.

Notice the size of the articles in relation to the 12-inch scale.

Turn to the next page.

New forging processes have been developed over the last 20 years. These processes attempt to combine the highest of standards while utilizing the minimum of material and energy resources. In fact, computer simulation is becoming an important tool in material behavior studies. Such activities can also provide forged products that are essentially ready to use right off the forge press.

Some of the newer forging processes include radial forging where dies approach the workpiece from radial directions. Precision forging produces parts of tight tolerances, usually made from the more exotic materials or when demands on clearances are tight. Powder forging is a heating and pressing operation where 100% theoretical density can be achieved. Materials that are not easily machined are candidates for such a process. Super plastic forging involves the use of a gas instead of hard punches to shape a material.

All of the above processes are designed to provide specific final product characteristics with emphasis placed upon economical considerations.

Turn to the next page for a short review.

CHAPTER REVIEW

____ 1. Forgings are pounded or pressed into shape from preheated metal. As the metal is forged, the grain _____ the shape of the dies.

    A. laps in
    B. folds into
    C. follows
    D. is independent of

____ 2. If the die happens to be misaligned, a _____ will be formed.

    A. burst
    B. surface discontinuity
    C. subsurface discontinuity
    D. crack

____ 3. The surface discontinuity is caused by folding of the metal in a thin plate. It is called a:

    A. forging lap.
    B. forged crack.
    C. burst.
    D. thin plate.

4. A forging lap:

    A. can be only half as large as the dies.
    B. is a subsurface discontinuity.
    C. is a crack.
    D. is always open to the surface.

5. Another possible cause of forging laps is an abrupt change in:

    A. grain direction.
    B. hot tops.
    C. bursts.
    D. seams.

6. Misaligned dies could be the cause of _____ which would be _____.

    A. seams, open to the surface
    B. laps, closed to the surface
    C. seams, closed to the surface
    D. laps, open to the surface

7. Laps are not the only discontinuity found in forgings. If the metal is _____ it might not flow properly, causing forging _____ or _____.

    A. underheated, bursts, cracks
    B. underheated, laps, seams
    C. overheated, bursts, cracks
    D. overheated, laps, seams

____ 8. Forging bursts or cracks are ruptures in a metal as it is being forced into a new shape. These discontinuities are:

   A. sometimes open for business.
   B. always subsurface.
   C. either surface or subsurface.
   D. always open to the surface.

____ 9. As one might expect, forging bursts or cracks are ragged in shape. These occur where the metal was:

   A. underheated.
   B. overheated.
   C. forged in mismated dies.
   D. lapped.

____ 10. If only the interior of the forging piece was underheated, the burst would:

   A. not be present.
   B. be found only under the surface.
   C. be found on the surface.
   D. be found following the grains.

Turn to the next page for answers to these review questions.

ANSWERS TO REVIEW QUESTIONS FOR CHAPTER 3

Question & Answer		Reference Page(s)
1.	C	3-5
2.	B	3-10
3.	A	3-12
4.	D	3-16
5.	A	3-13
6.	D	3-14
7.	A	3-22
8.	C	3-22
9.	A	3-22, 3-28
10.	B	3-26

For an account of parts made from molten metal, turn to Chapter 4.

CHAPTER 4

CASTING DISCONTINUITIES

Castings are made by pouring liquid metal into a mold. These molds are formed close to the shape of the finished part. For example, many people cast their own bullets by melting lead and pouring it into a mold. When the metal solidifies, it is removed from the mold.

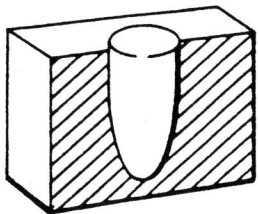 

Casting processes can be divided into two major groups: permanent patterns and expendable patterns. Permanent mold castings use reusable dies into which the molten metal is poured. Once cooled, the die is separated and the workpiece is removed.

Investment casting is an expendable cast process which involves the application of a ceramic slurry around a disposable pattern, usually wax. After the slurry hardens, the wax is melted out of the mold, then the molten metal is poured into the cavity. The mold is destroyed to reveal the cast workpiece.

Turn to the next page.

There are several other casting processes, a couple of which are described briefly at the end of the chapter. We will now focus on green sand mold casting, which is an expendable casting process.

Casting molds are usually made from sand, clay, and water. The clay and water form a thin film over each granule of sand, joining the granules to make the mold. As you might suppose, this mold can be easily broken and is permeable (absorbent). These qualities are exactly what is required in a mold. You might also have guessed that castings have no regular grain structure. There is no rolling or forging to give direction to the grain.

Turn to the next page.

The illustration below shows a casting being poured. When the casting solidifies, it will have an irregular grain structure.

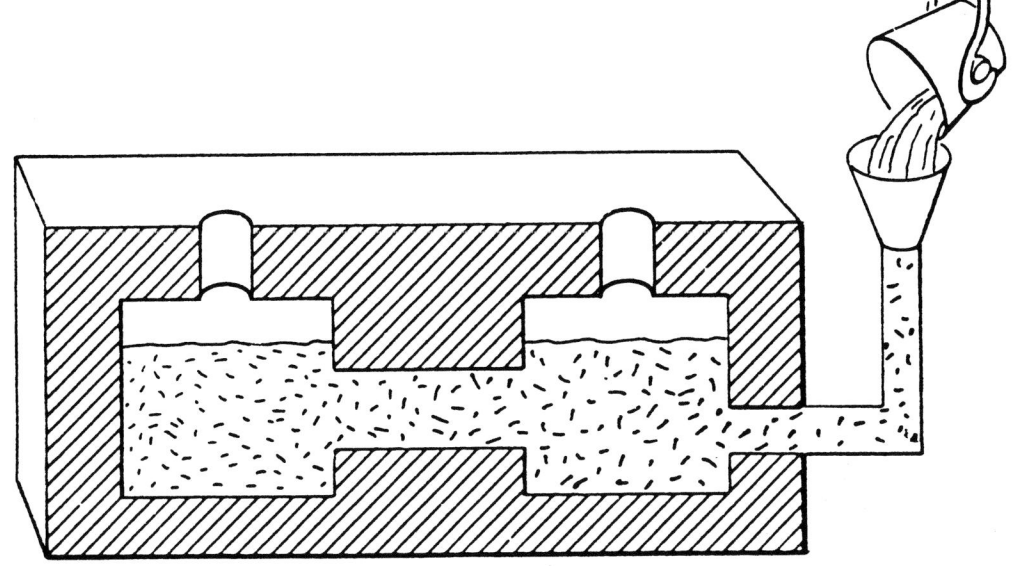

Turn to the next page.

Casting discontinuities can be identified by a diverse array of names. Foundrymen have coined many clichés for casting discontinuities over the years. *The International Committee of Foundry Technical Associations* has divided casting defects into several basic categories. These include metallic projections, cavities, inclusions, incomplete casting, incorrect dimension and discontinuities to name a few.

We are going to discuss several common types of discontinuities found in castings. The first discontinuity is a COLD SHUT.

Turn to the next page.

From page 4-4

COLD SHUTS are caused when molten metal is poured over solidified metal. When the metal is poured, it hits the mold too hard and spatters small drops of metal. When these drops of metal hit higher up on the mold, they stick and solidify. When the rising molten metal reaches and covers the solidified drops of metal, a crack-like discontinuity is formed.

When are cold shuts formed?

When molten metal is poured over solidified metal . . . . . . Page 4-6
When molten metal is poured over molten metal . . . . Page 4-8

Good. A COLD SHUT is formed when molten metal is poured over solidified metal.

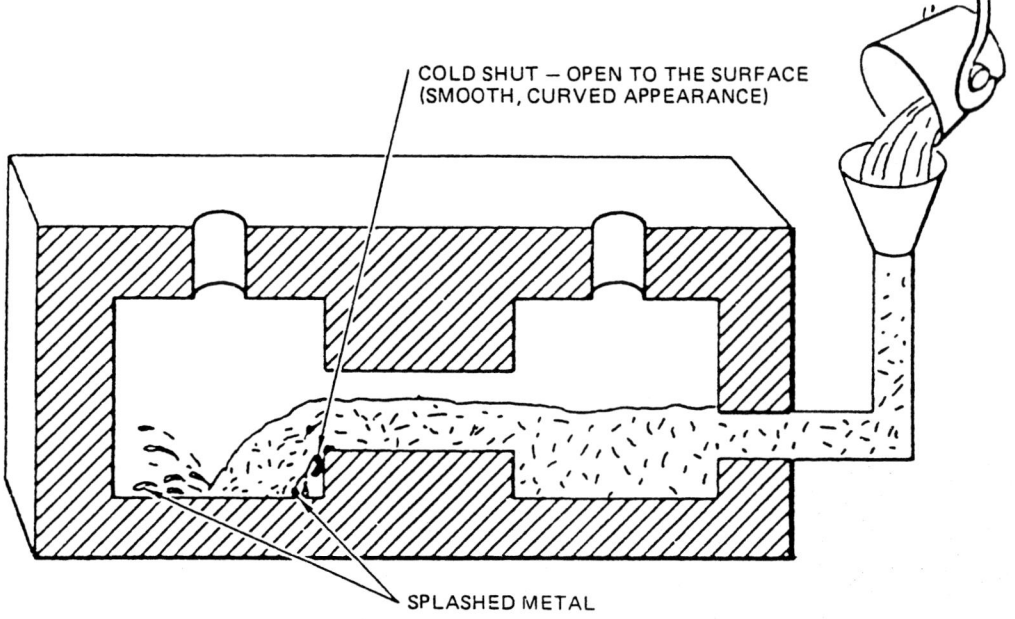

Cold shuts can also be formed by the lack of fusion between two intercepting surfaces of molten material of different temperatures.

Cold shuts are one type of discontinuity. The second type of discontinuity found in castings is HOT TEARS (shrink cracks). As the name implies, these cracks result from a tearing action within the metal. To better understand why this tearing action occurs, let's take another look at the ingot stage of steel-making.

Turn to the next page.

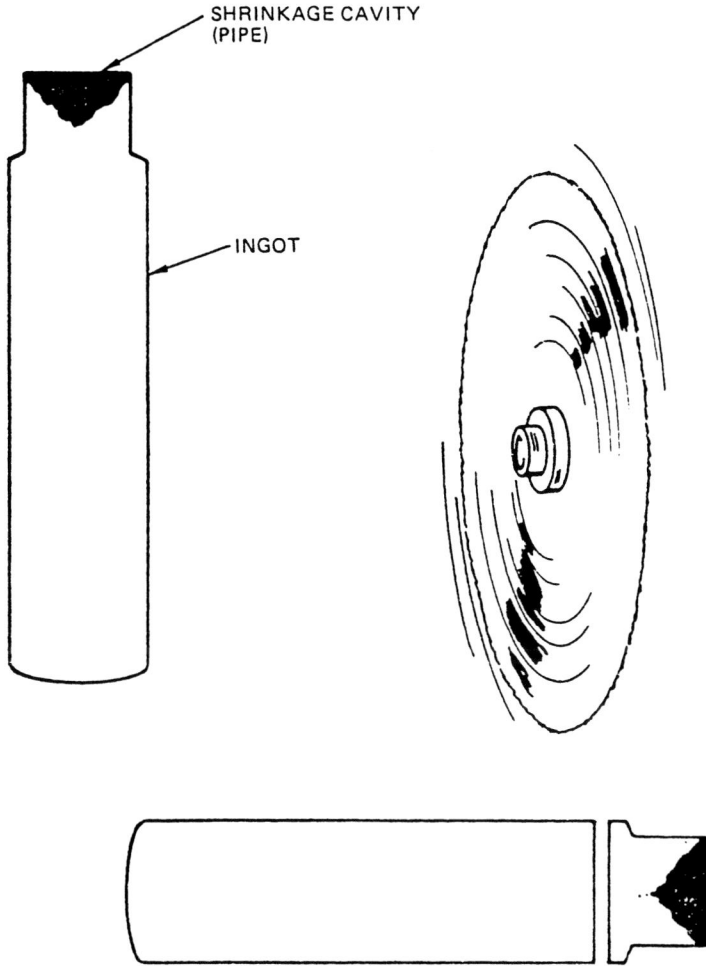

The illustrations above should recall to mind that molten metal occupies more space than solid metal. In other words, metal shrinks as it solidifies. As the ingot solidifies, a shrinkage cavity forms in the top. The ingot is cropped, eliminating the cavity.

Turn ahead to page 4-9.

From page 4-5

Molten metal poured over more molten metal would not form a cold shut. The clue here is in the word COLD. A cold shut can only be formed by molten metal meeting with metal that has solidified or is relatively *cold*.

Turn back to page 4-6 and continue.

The top of the open ingot was a natural place for shrinkage to occur. In a casting, the shrinkage problem is not so simple. Consider the illustration below which shows hot tears caused by shrinkage.

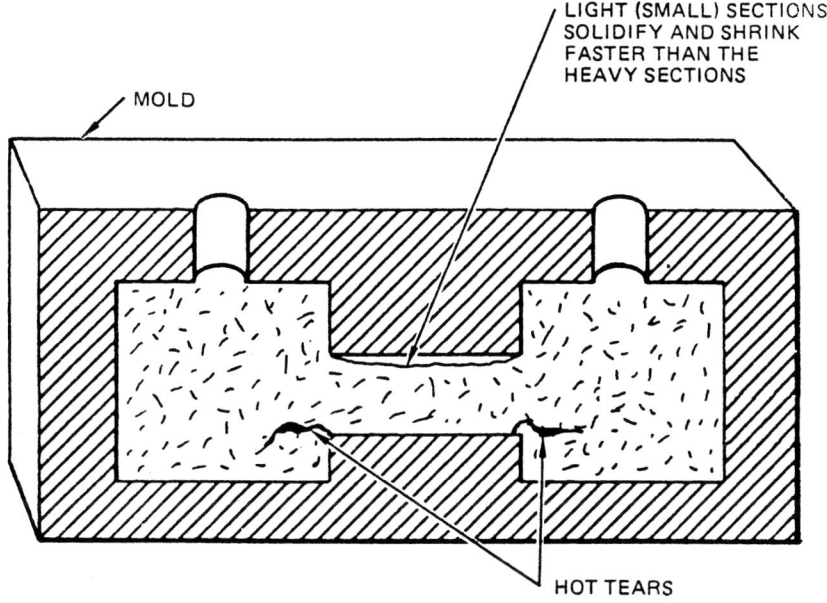

In a casting having light and heavy sections, the light sections, being smaller, solidify faster; they shrink faster pulling the heavier sections, which are hotter and not shrinking as fast, toward them.

This unequal shrinking between the light and heavy sections can cause:

shrinkage depression .... Page 4-10
hot tears .... Page 4-12

From page 4-9   4-10

Don't misplace the shrinkage depression - it belongs in the ingot! While the same action (metal shrinkage) causes shrinkage depression in an ingot and hot tears in a casting, a casting is more complicated in shape than an ingot, and this force can cause more trouble than mere shrinkage depression.

Turn back to page 4-9 and reread the question.

From page 4-12   4-11

Excellent. Hot tears are most likely to occur at junctions of light and heavy sections of a casting. Let's take a look at a casting removed from its mold.

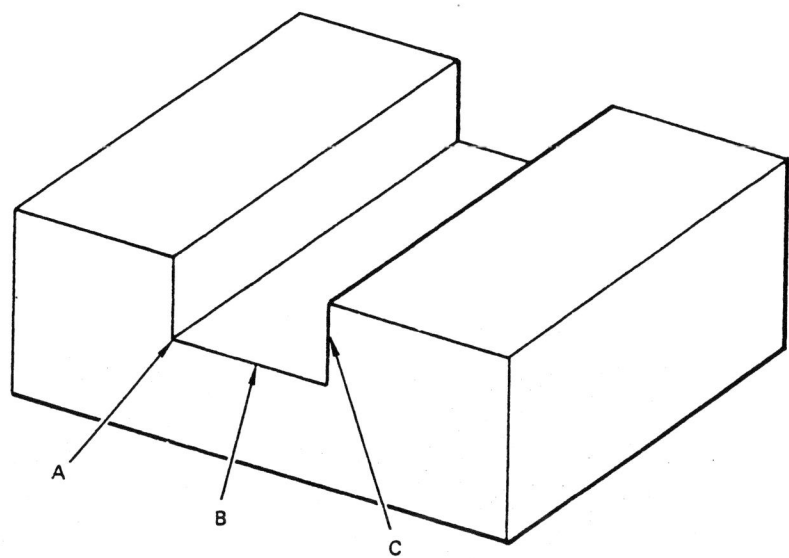

Where do you think a hot tear (shrink crack) would most likely occur in this casting?

Point A .... Page 4-13
Point B .... Page 4-15
Point C .... Page 4-17

Well chosen! If a casting has light sections and heavy sections, the light sections solidify faster and shrink faster than the heavy sections. This unequal solidifying causes the faster-shrinking light sections to apply a pulling stress on the heavier sections. This creates stresses that *can* result in HOT TEARS *at the junction of the light and heavy sections.*

These tears can be prevented. In our discussion of castings, the composition of the mold was mentioned: sand, held together by clay and water. The mold is easily broken and is permeable (absorbent).

Hot tears can be prevented if, when stresses are built up by unequal shrinkage, the mold breaks before the metal tears. This breaking of the mold, as the heavier sections are pulled against it, takes up the stress that might otherwise tear the metal.

Where are hot tears most likely to occur in a casting?

At the junction of light and heavy sections . . . . Page 4-11
Where the heavy sections contact the mold . . . . Page 4-14

Correct. You might possibly find a hot tear at point A.

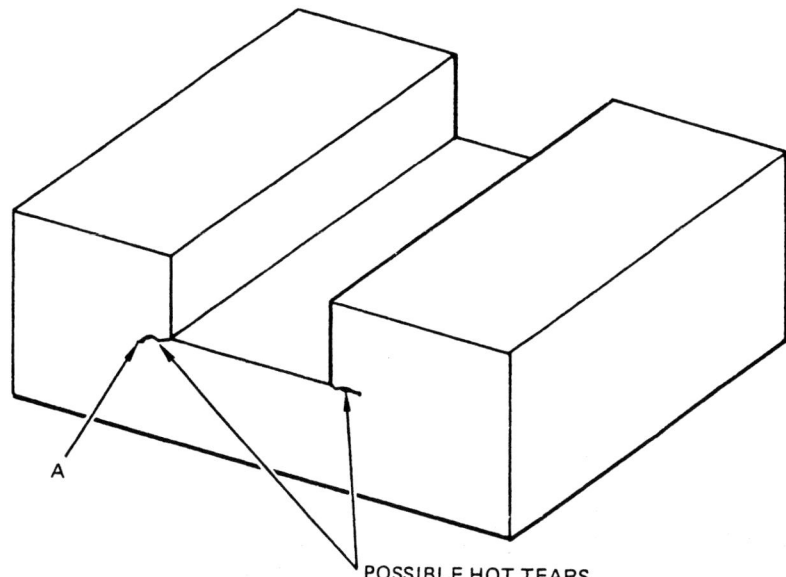
POSSIBLE HOT TEARS

Points B and C were not located at the junction between heavy and light sections of the casting, but you might find another tear on the right side at the other junction of the light and heavy sections, as shown above.

Turn ahead to page 4-16.

From page 4-12   4-14

"Where the heavy sections contact the mold" is not the correct answer to the question:

"Where are hot tears most likely to occur in a casting?"

Remember that the name "hot tear" is descriptive. At a junction of light and heavy sections, unequal solidifying may cause the solidified, light section to *tear* away from the heavier, hot section.

Turn back to page 4-11 and continue.

Your answer, point B, indicates a belief that a hot tear might be likely to occur here. While this area is the light section, point B is not a *junction* of the light and heavy sections. A hot tear would not be apt to occur here.

Turn back to page 4-11, take a second look at the picture and select another answer.

From page 4-13   4-16

Hot tears do not look like COLD SHUTS. Hot tears have a ragged crack-like appearance.

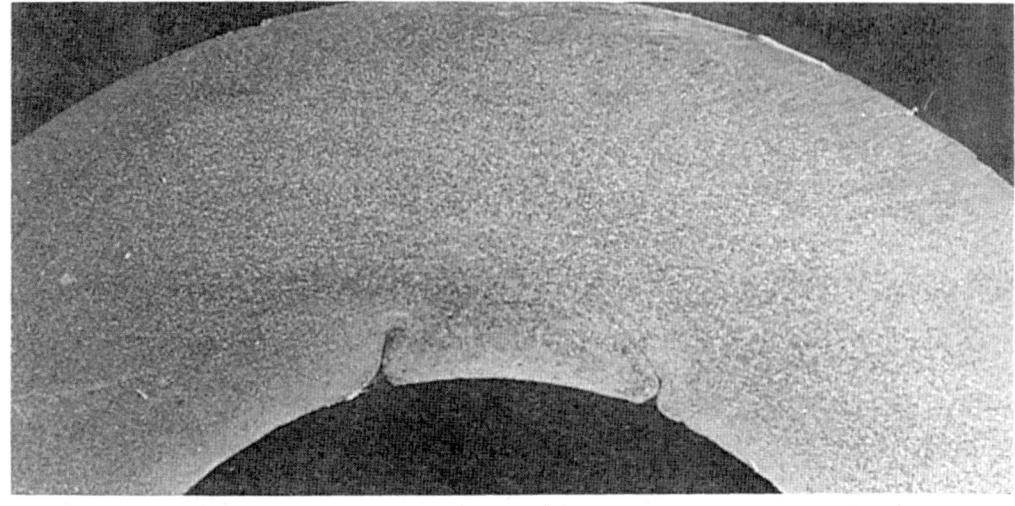

What kind of a discontinuity is shown in this casting?

Hot tear . . . . **Page 4-18**
Cold shut . . . . **Page 4-21**

From page 4-11

You chose point C as the location at which a hot tear would most likely have occurred. Point C is in a heavy section, but it is not a junction between a light and a heavy section. It is at these *junctions* that a hot tear is most likely to occur.

Turn back to page 4-11 for another look at the picture and select the correct answer.

From page 4-16

The discontinuity shown was not a hot tear. The discontinuity was pictured as a curving, not ragged, line. A hot tear was described as having a ragged appearance.

Turn ahead to page 4-21 for another word about cold shuts.

No. The discontinuity shown is not a hot tear. A hot tear is caused by unequal shrinking of light and heavy sections of a casting as the metal cools. Here is the way hot tears look.

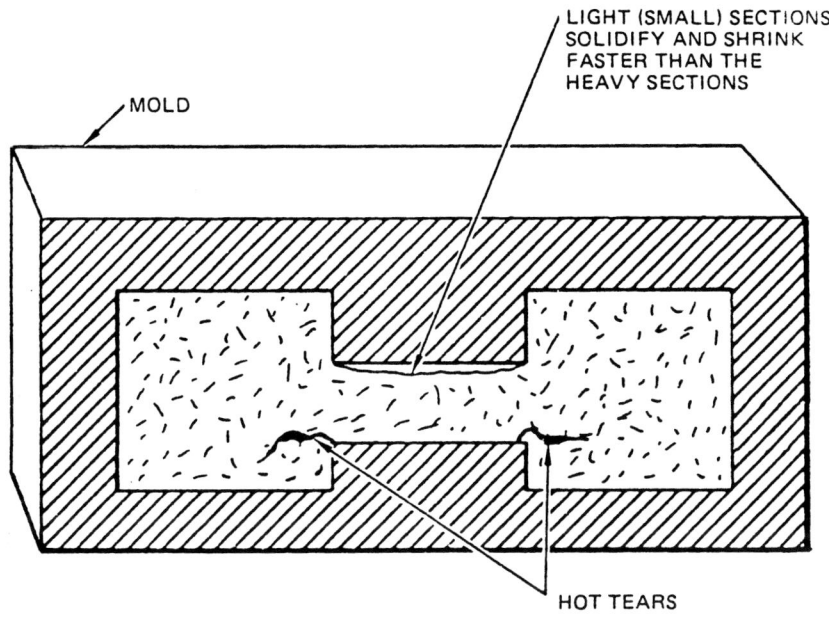

The discontinuity in question is one caused by intersecting surfaces of molten metal at different temperatures.

Turn ahead to page 4-21, review the photos, and select the other answer.

Right. Those were hot tears - shrink cracks. Here is a close-up of one of those shrink cracks.

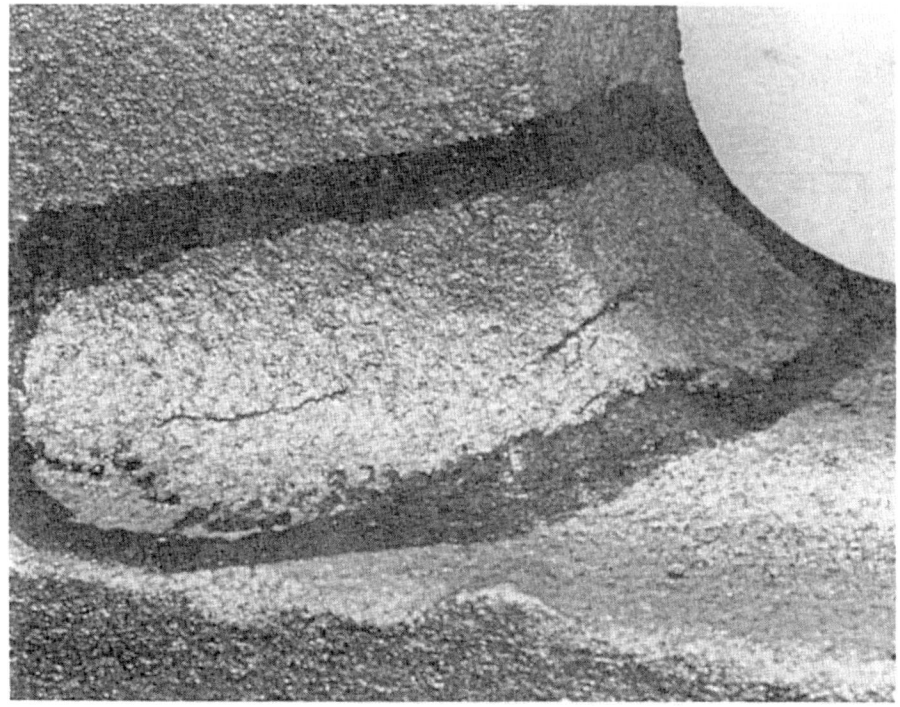

Notice the coarse texture of the surface of the casting and the ragged nature of the cracks.

Turn ahead to page 4-22.

From page 4-16

4-21

That's a cold shut all right. A hot tear looks like a crack. Fine. Here is a cast aluminum joint.

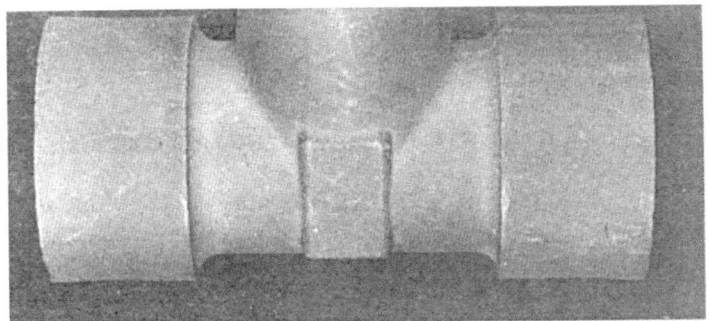

Inside of the joint we have a discontinuity as shown by the following cross-sectional views.

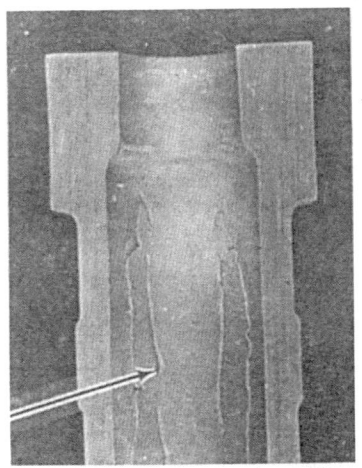
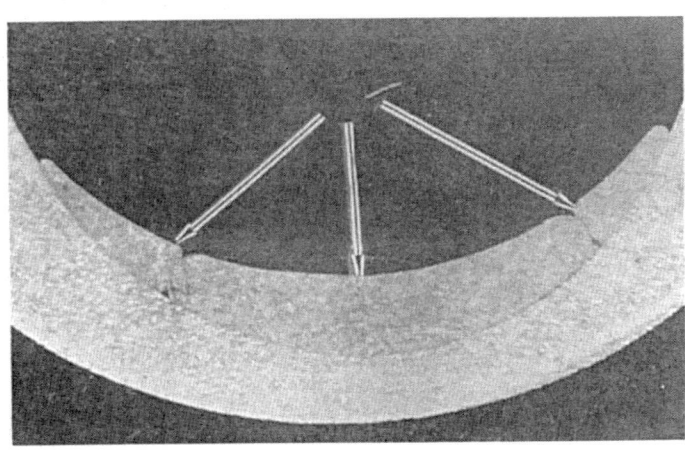

When the casting was poured, intersecting surfaces of the molten metal at different temperatures did not fuse on cooling.

What is the name of this discontinuity?

Hot tear (shrink crack) .... **Page 4-19**
Cold shut .... **Page 4-23**

Here is another example of a hot tear or shrink crack. Here again, notice the coarse texture of the casting.

Below is a close-up of the shrink crack.

Another possible type of discontinuity in castings, and our next topic of discussion, is a SHRINKAGE CAVITY.

Turn ahead to page 4-24.

From page 4-21   4-23

Absolutely. That was a cold shut. When the casting was poured, some of the molten metal splashed onto the mold wall and solidified. When the rising molten metal reached the solid metal it did not fuse and a smooth, curving COLD SHUT was formed.

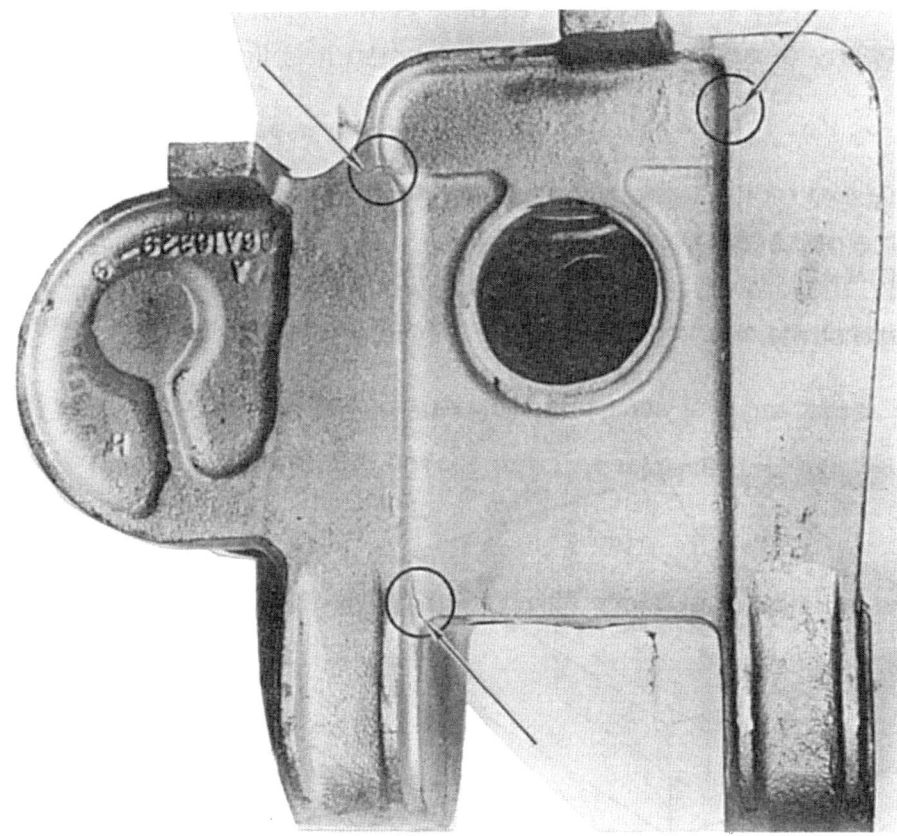

Here is another casting with ragged-looking discontinuities pointed out by the arrows. What is the name of this type of discontinuity?

Hot tear (shrink crack) .... **Page 4-20**
Cold shut .... **Page 4-25**

From page 4-22

4-24

Shrinkage Cavities

Shrinkage cavities occur when there is insufficient molten metal to fill the space created by shrinkage, just as pipe is formed in an ingot. Metal has this property of occupying more space when it is liquid than when it is solid.

Here are two molds. The dotted lines indicate the shape of the casting inside the mold. The drawing on the left depicts the metal while it is still molten. The drawing on the right shows the solidified casting with a shrinkage cavity.

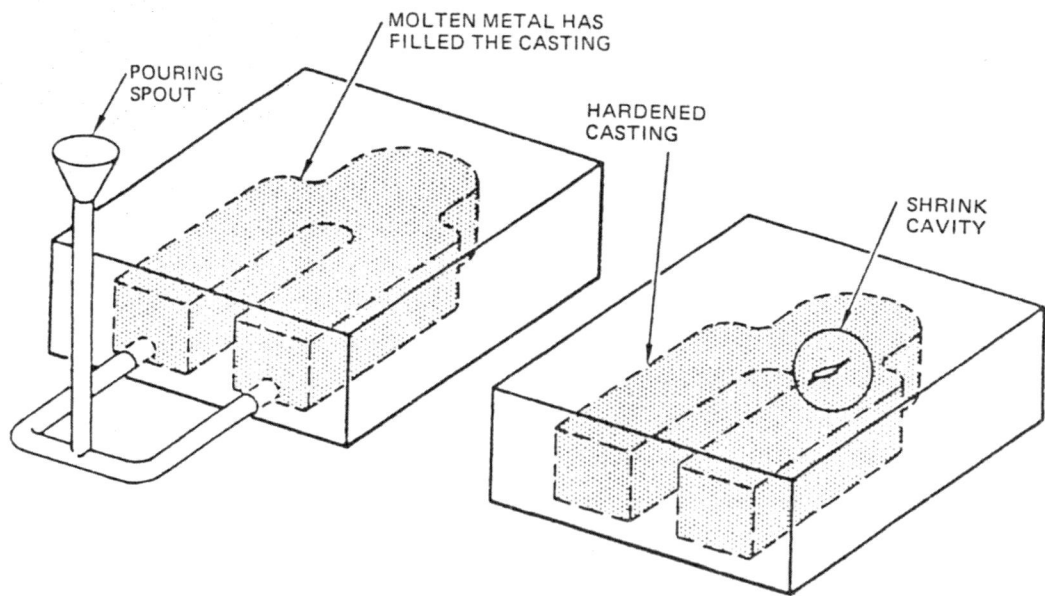

Why is there a cavity in the solidified casting?

**Only enough metal was poured to exactly fill the mold while
the metal was liquid** .... **Page 4-27**
The metal burst in the center as it solidified .... **Page 4-28**

The discontinuity shown was not a cold shut. It did not show points where the metal did not fuse because of splashes on the mold wall; instead it showed tears or cracks in the casting. These are *hot tears* (shrink cracks).

For a closer look at a hot tear, turn back to page 4-20.

From page 4-27   4-26

Shrinkage cavities can be eliminated, or the possibility of shrinkage cavities can be greatly reduced, by adding a feeder head or reservoir as shown below.

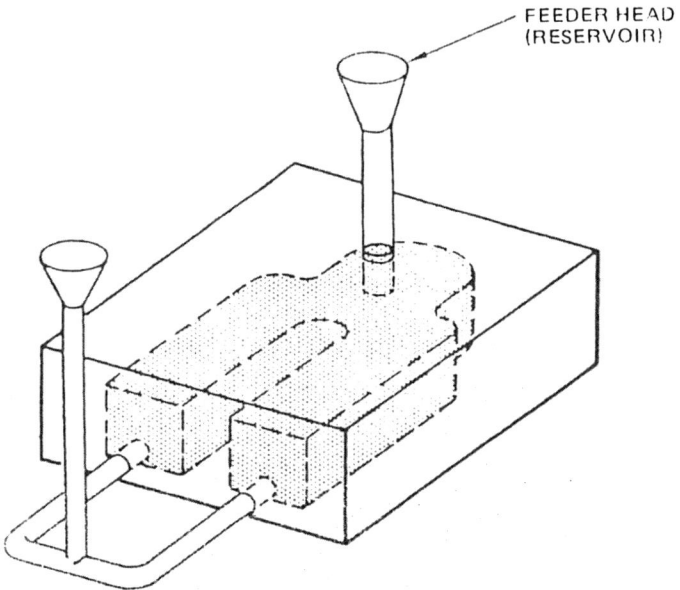

Except for the addition of the feeder head, the casting above is exactly the same as the one shown on page 4-24.

The picture shows that more metal was poured into this casting than into the previous castings. After the casting was filled, the extra metal rose into the feeder head.

The additional metal, which fills the feeder head, will feed back into the casting to replace the shrinking metal. Under these conditions shrinkage will:

eventually become a small, harmless cavity within the casting .... Page 4-29
be absorbed by feeder head excess .... Page 4-30

From page 4-24

Yes, there is a cavity because only enough metal was poured to exactly fill the mold when the metal was liquid. The metal solidified from the outer surface of the casting toward the heaviest part of the casting. As it was solidifying, the metal was also shrinking, and the solidified portion was pulling away from the liquid portion. Finally, as the last of the metal solidified and shrank away, a space remained because there was no metal left to compensate for the shrinkage.

Turn back to page 4-26.

From page 4-24   4-28

The cavity in the casting shown on page 4-24 was not caused by bursting of the metal as it solidified. A burst is a discontinuity caused by forging processes.

The cavity in the casting was caused by shrinkage of the metal as it solidified. For this reason, it is called a shrinkage cavity.

Turn back to page 4-27 and continue.

From page 4-26

No, the purpose of the feeder head is to help eliminate shrinkage cavities.

Perhaps thinking of the process as a series of steps will help to establish it in your mind.

- The metal is poured into the casting and fills the casting.

- Pouring of the metal continues and the feeder head is filled.

- A thin layer of the casting metal begins to cool and shrink.

- Metal from the feeder head immediately replaces the space created by that shrinkage.

- Another thin layer of the casting cools and shrinks.

- Again metal from the header feeds in to replace the space created by shrinkage.

This series of steps is repeated again and again until finally the last metal to cool and shrink is the metal in the head. No metal is left to fill in there, but it doesn't matter. The feeder head has fulfilled its purpose and will be removed.

Turn to the next page.

Correct. The shrinkage will be absorbed by the excess metal in the feeder head.

Since casting design can take any number of shapes, shrinkage cavities can appear at many points in a casting. Such is the case in this stainless steel casting.

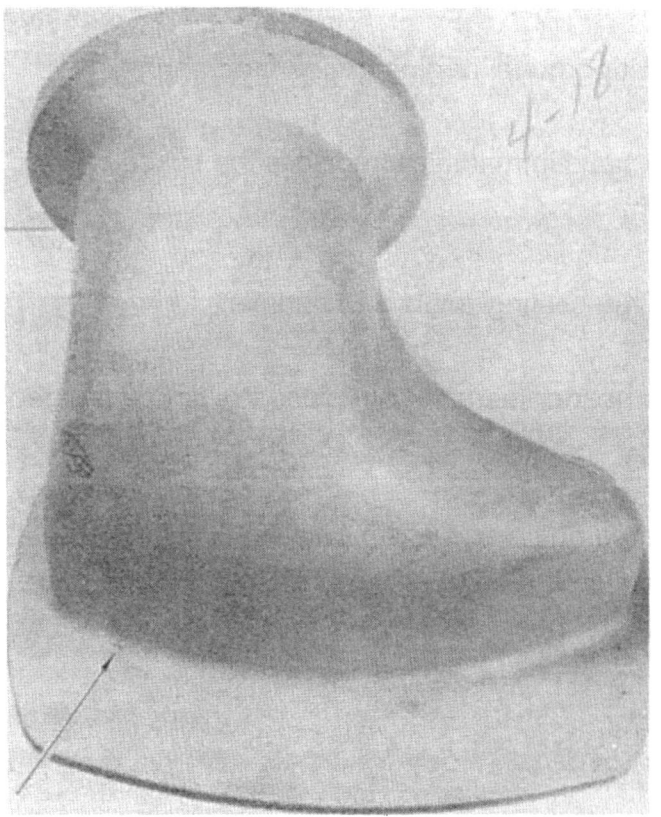
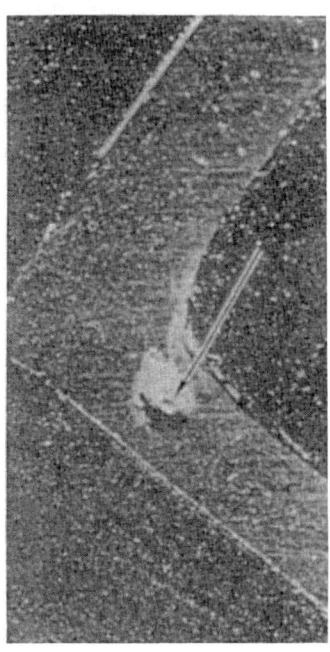

The shrinkage cavity in this casting has a very small surface opening, which is hardly visible. When the casting is cut through this section, the shrinkage cavity is clearly visible, as shown by the picture on the right.

Turn to the next page.

Shrinkage can also occur in the casting at the mold gate - the entrance to the mold through which the molten metal is poured.

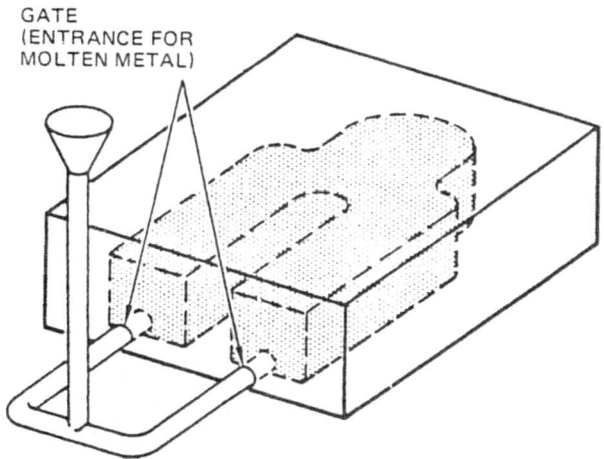

Shrinkage can occur if metal at the gate solidifies or is blocked off while some of the metal beneath is still molten. Shrinkage which occurs at the gate appears as many small holes called MICROSHRINKAGE (small shrinkage). Since microshrinkage is caused by premature blocking of the gate, it is subsurface.

Turn to the next page.

From page 4-31   4-32

Microshrinkage can also occur deeper within the metal if the mold is improperly designed. For instance, metal might be poured into a heavy section of mold, flow through a light section, then into another heavy section of the mold. The entrance from the light into the heavy section might block off prematurely just as the gate did. If this happens, then microshrinkage would occur deeper within the metal.

If you were now asked to inspect a casting, where would you first look for possible discontinuities?

At junctions between light and heavy sections and at the gate .... **Page 4-35**
Along the edges of the casting .... **Page 4-36**

From page 4-39

4-33

You would expect to find blow holes somewhere on the surface of the casting, not at the gate where the metal was poured.

Blow holes are caused from steam within the mold. They would occur where the casting contacts the mold - the casting's surface.

Turn to the next page.

From page 4-39

Yes. Since blow holes are caused by steam within the mold, these discontinuities would occur somewhere on the surface of the casting - the point of *contact with the mold*.

The same principle applies in the case of cores, although the core is held together by oil rather than water. A CORE is used to occupy the internal spaces or voids in a casting mold. Since a core is surrounded by metal, the core must be vented to release the oil vapors. Inadequate venting can force the vapors into the surface of the casting where the casting contacts the mold core.

Turn ahead to page 4-40.

From page 4-32   4-35

Of course. If you were given a casting, the most logical place to look for discontinuities (on the basis of information thus far) would be at junctions between light and heavy sections and at the entrance to the gate.

One discontinuity that might occur at the junction between light and heavy sections is called a *hot tear*. It would be due to shrinkage.

Another discontinuity that might occur at light and heavy junctions or at the gate is also due to shrinkage and is called:

porosity . . . . **Page 4-37**
microshrinkage . . . . **Page 4-38**

From page 4-32

If given a casting for inspection, the most logical place to look for discontinuities, on the basis of information thus far, *would not* be along the edges of the casting.

We have discussed premature blocking of entrances, either at the gate or at the junction of a light and heavy section of a casting. Logically then, you would look for discontinuities at these two locations.

Turn back to page 4-35 and continue.

From page 4-35

The name of the discontinuity we want is MICROSHRINKAGE, *not* porosity. Although MICROSHRINKAGE looks somewhat like porosity, the clue to the name of the discontinuity is in the word *shrinkage*. Premature blocking of openings can cause still-molten metal beneath the opening to shrink, creating an area of small holes. Micro means small, so microshrinkage is the term for this discontinuity.

Turn to the next page and continue.

From page 4-35

Correct. Microshrinkage is a discontinuity that might be found at junctions of light and heavy sections or that might be found at the gate. It is caused by premature blocking of an entrance either near the surface or deeper within the metal. Microshrinkage, therefore, is subsurface.

Turn to the next page for a discussion of still another possible discontinuity in castings - BLOW HOLES.

Blow Holes

Blow holes are small holes in the surface of the casting caused by gas that is not within the molten metal - external gas. This external gas comes from the mold itself.

Remember the makeup of the mold? Its composition is sand, clay, and water, which is permeable or absorbent. When the molten metal contacts the mold, steam is formed by the water which is part of the mold. If the mold is permeable enough, the steam is forced through the mold to the outside.

If the mold is not permeable enough, the steam cannot get through to the outside. Since it must go somewhere, it is forced back into the casting, blowing holes in the casting's surface - BLOW HOLES.

Where might you expect to find blow holes in a casting?

At the gate where the metal was poured . . . . Page 4-33
Somewhere on the surface of the casting . . . . Page 4-34

From page 4-34   4-40

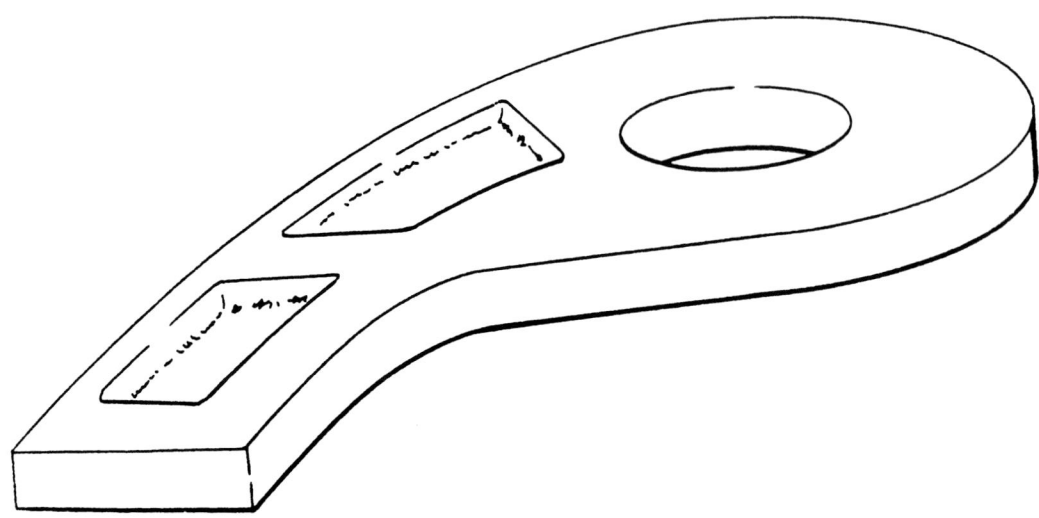

From the illustration of the casting above, where do you think blow holes might occur?

On the surface of the casting but not in the wall of the hole .... Page 4-43
On the surface of the casting and on the wall of the hole .... Page 4-44

We have discussed six types of discontinuities that might be found in or on the surface of castings. The names of the discontinuities provide a clue to their cause.

- COLD SHUTS - Hot metal over solidified metal or intersecting surfaces at different temperatures.

- HOT TEARS (SHRINKAGE CRACKS) - Tears in metal from unequal cooling.

- SHRINKAGE CAVITIES - Subsurface cavities caused by shrinkage.

- MICROSHRINKAGE - Many small, subsurface cracks or cavities.

- BLOW HOLES - Holes blown into a casting's surface.

- POROSITY - Entrapped gas.

One should be aware that it is possible that more than one type of discontinuity may be present on or in a casting. The presence of these discontinuities may also provide useful information to the foundrymen as to their cause and remedy.

Turn to the next page.

Newer casting processes have been developed which primarily include some form of counter-gravity process. This process draws the molten metal up into the mold against gravity, as the name implies. These techniques allow us to make parts with thin sections and intricate detail. Additionally, this process allows us the flexibility to produce parts from metals that are normally reactive to air by counter-gravity techniques in a vacuum.

Casting techniques are now available that allow monocrystal solidification. That is, we can readily produce a turbine blade that is a single crystal! Such a turbine blade is advantageous when subjected to the combined temperatures and stresses in a jet engine.

Semisolid casting is a combined process of both casting and forging. The metal billet is cut to the appropriate weight, heated to a semisolid state, and then pressed into shape in a set of dies. The thrust of this effort, beginning in the United States in the early 1970's, was to control the grain structure of the final product. Automotive brake master cylinders, aluminum wheels, plumbing fittings, and electrical connectors have all been made in this manner.

Before going on to other metal-forming processes, turn ahead to page 4-45 for some examples of space-age-type precision castings. Notice the intricate design and small size of these castings in relation to the scale at the bottom.

From page 4-40

You answered that blow holes might be found on casting surfaces but would not be found on the walls of the hole. That answer is incorrect because the hole walls were formed by part of the mold - the core. So blow holes might be found on casting surfaces as well as the hole walls, which are actually part of a casting's surface.

Turn to the next page.

From page 4-40

Right. Blow holes can occur on the surface of the casting as well as on the wall of the hole. Both of these surfaces touch the mold.

Still another discontinuity found in castings is the now familiar POROSITY. Porosity in castings is caused by entrapped gas - the same way it is caused in the ingot. Porosity can be either at the surface or subsurface, depending on the design of the mold or configuration of the article. Here is an example of porosity in the surface of a casting.

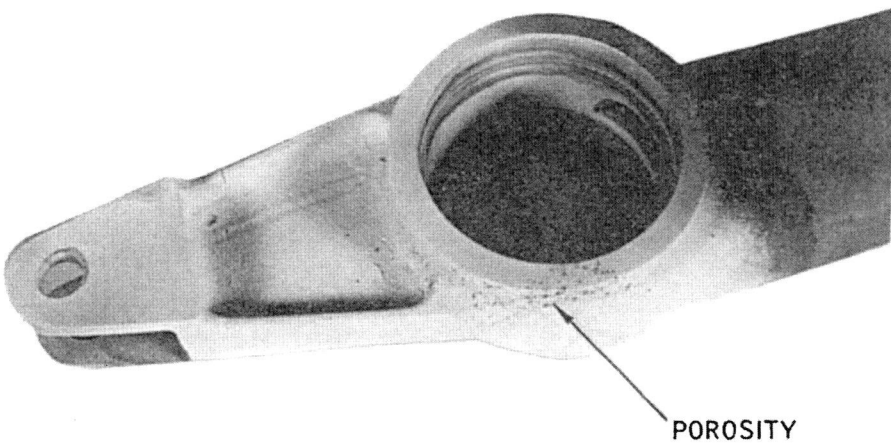

Turn back to page 4-41.

From page 4-42 4-45

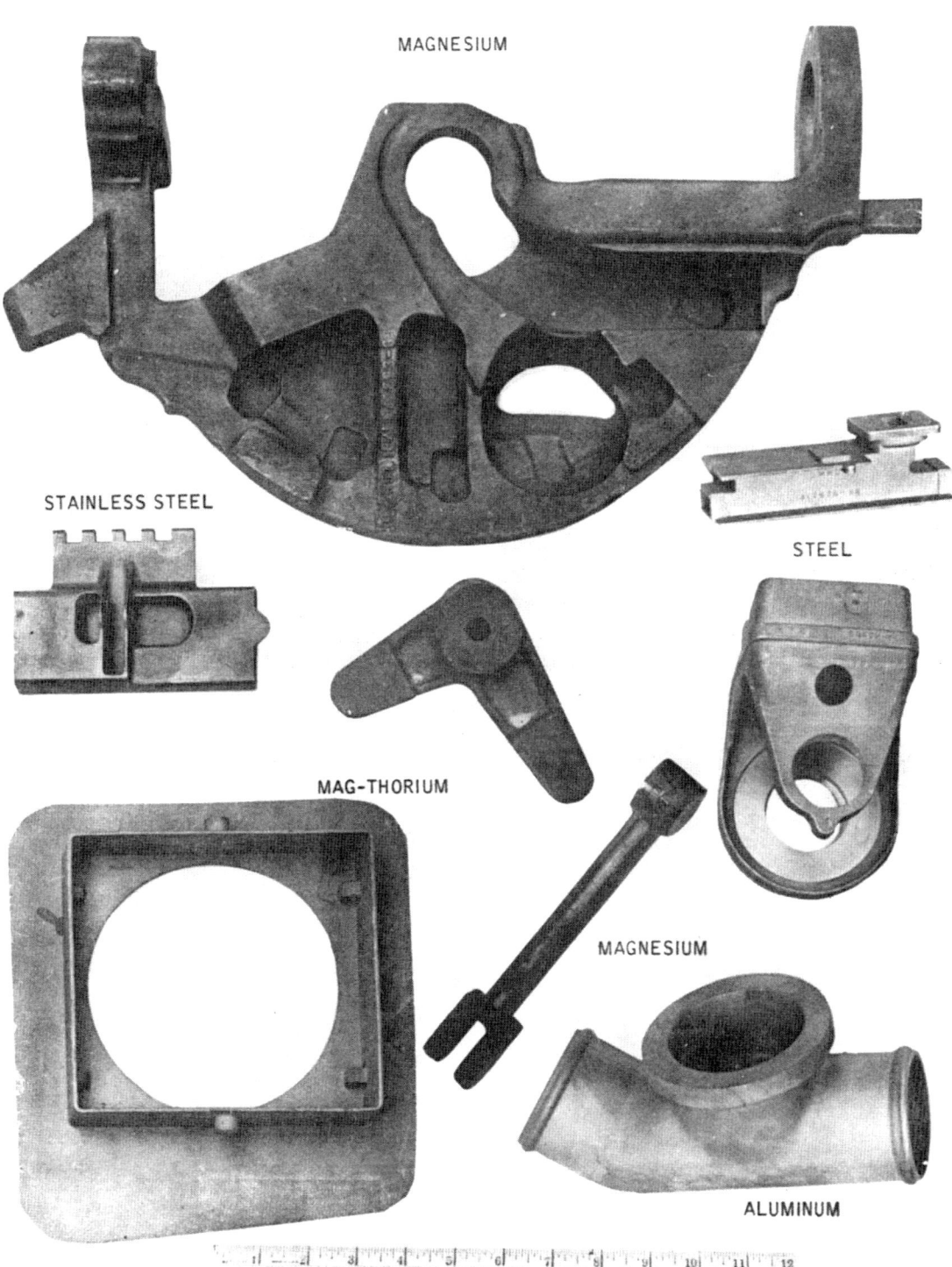

Turn to the next page.

These castings are all steel with variations in alloys.

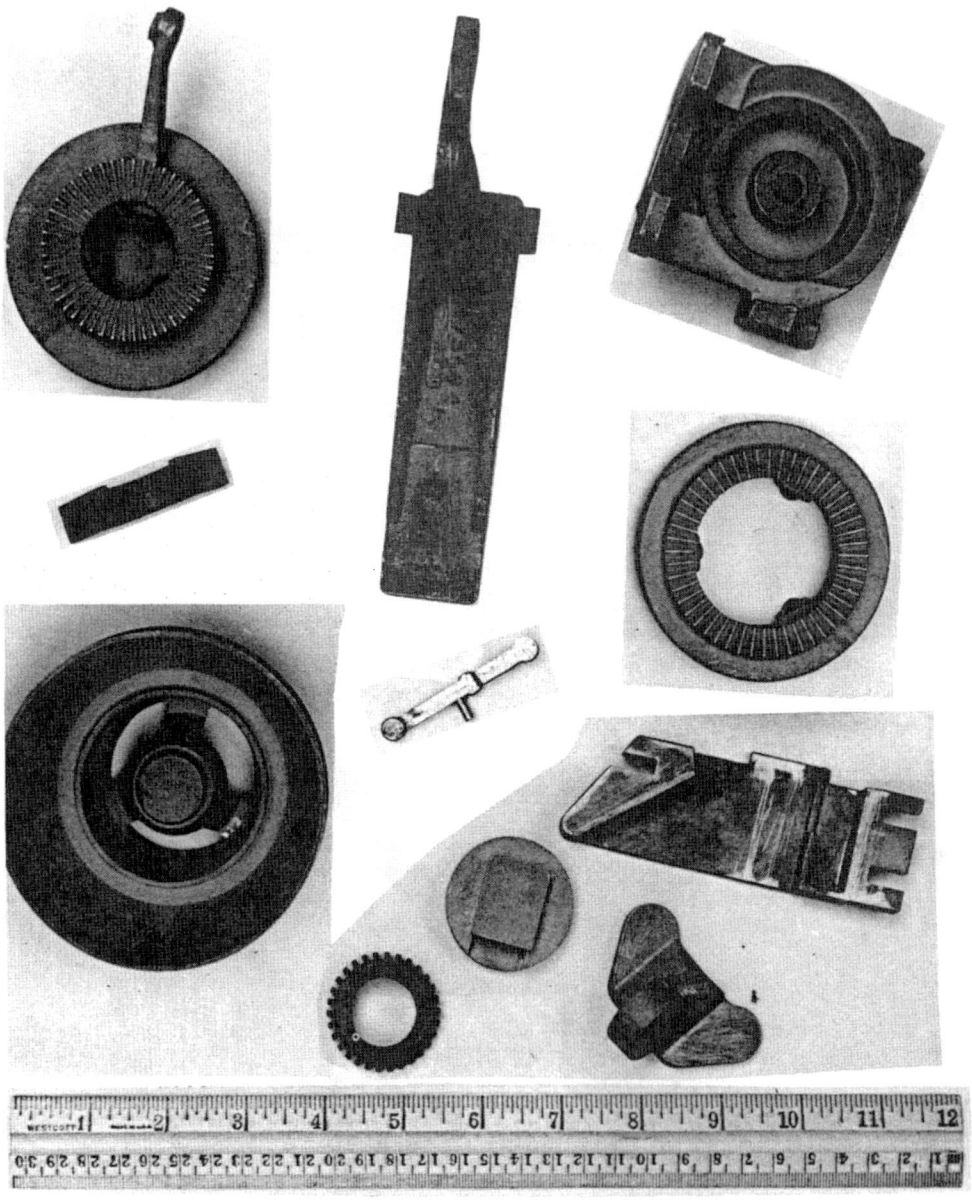

Turn to the next page for a review of casting discontinuities.

CHAPTER REVIEW

1. Castings are made by pouring liquid metal into a mold. Several kinds of breaks or _____ can occur in castings.

    A. folds
    B. laps
    C. cracks
    D. discontinuities

2. When the casting is poured, drops of metal can splash up on the inside walls of the mold. These drops stick and solidify. When the rising metal covers these solidified drops, crack-like discontinuities are formed. These are called:

    A. cold shuts.
    B. drop shuts.
    C. laps.
    D. hot tears.

3. Cold shuts within a casting, which are a result of molten metal covering solidified metal or two intersecting molten surfaces at different temperatures, have a smooth, _____ appearance.

    A. straight
    B. slightly curved
    C. diagonal
    D. blow hole

4. A smooth, curved discontinuity in a casting might indicate a:

   A. hot tear.
   B. cold shut.
   C. microshrinkage.
   D. blow hole.

5. A discontinuity with a ragged, crack-like appearance and which results from metal shrinkage is called a:

   A. hot tear.
   B. cold shut.
   C. blow hole.
   D. hot shut.

6. Light (thin) sections next to heavy (thicker) sections in castings cause different cooling rates that can result in _____ of the cast metal.

   A. tearing
   B. collapsing
   C. microshrinking
   D. blocking

7. Hot tears are most likely to occur at the junctions of _____ and _____ sections.

   A. round, long
   B. short, long
   C. round, short
   D. thick, thin

8. Cold shuts have a smooth, curved appearance. Hot tears are _____ in appearance.

   A. smooth and curved
   B. smooth and straight
   C. smooth and run in multiple directions
   D. ragged

9. Another discontinuity that can develop in castings is a result of the property that metal has of _____ as it solidifies.

   A. expanding
   B. shrinking
   C. blocking
   D. gating

10. The discontinuities associated with shrinkage are called:

    A. cast expansion.
    B. hot tears.
    C. blocking.
    D. gating.

11. Shrinkage cavities are caused by poor mold design in that not enough metal is poured into the casting to allow for shrinkage. Often, the metal shrinks and ___ _____ are formed.

    A. surface blocks
    B. subsurface cavities
    C. blocking cavities
    D. gates

12. The addition of a feeder head to feed additional molten metal into the casting will prevent:

    A. gates.
    B. hot tears.
    C. shrinkage cavities.
    D. cold shuts.

13. Shrinkage can cause another discontinuity that would also be located beneath the metal's surface. This shrinkage is located at the _____ where the metal enters the casting.

    A. shrinkage
    B. tear
    C. gate
    D. shut

From page 4-50

14. Unlike a shrinkage cavity, shrinkage at the gate is not a single cavity. Shrinkage that occurs at the gate takes the shape of many small holes called:

   A. microshrinkage.
   B. block holes.
   C. porosity.
   D. nonmetallic inclusions.

15. Microshrinkage is actually small shrinkage, and is caused by premature:

   A. cold shuts.
   B. gate porosity.
   C. blow holes.
   D. gate blocking.

16. The gate where the metal is poured is not the only place where premature blocking or shutting off of an entrance might occur. Another possibility is where _____ and _____ sections meet.

   A. round, long
   B. short, fat
   C. thick, thin
   D. round, short

| From page 4-51 | 4-52 |

___ 17. The junction between section thickness changes could act as an entrance, prematurely block off, and cause:

    A.   microshrinkage.
    B.   hot tears.
    C.   cold shuts.
    D.   hot shuts.

___ 18. A mold is made up of mostly sand, clay, and water. When the molten metal is poured into the mold the water boils, releasing steam, which is _____ if the mold is made correctly.

    A.   absorbed by the molten metal
    B.   blocked by the gate
    C.   released by the gate
    D.   vented to the outside

___ 19. If the mold is not made properly, the steam will be forced:

    A.   out of the mold and cause cold shuts.
    B.   out of the mold and cause blow holes.
    C.   into the casting surface causing shrinkage.
    D.   into the casting surface causing blow holes.

From page 4-52

20. Because blow holes can occur anywhere the mold contacts the casting, they are:

    A. subsurface discontinuities.
    B. surface discontinuities.
    C. similar to shrinkage.
    D. next to gates.

21. Gas entrapped in a casting is called _____ and is _____.

    A. porosity, always subsurface
    B. porosity, always surface
    C. porosity, either surface or subsurface
    D. blow holes, either surface or subsurface

For the correct answers, turn to the next page.

From page 4-53 4-54

ANSWERS TO REVIEW QUESTIONS FOR CHAPTER 4

Question & Answer		Reference Page(s)
1.	D	4-4
2.	A	4-5
3.	B	4-16
4.	B	4-16
5.	A	4-19
6.	A	4-12
7.	D	4-12
8.	D	4-19
9.	B	4-9
10.	B	4-9
11.	B	4-41
12.	C	4-26
13.	C	4-31
14.	A	4-31
15.	D	4-31
16.	C	4-32
17.	A	4-32
18.	D	4-34
19.	D	4-33
20.	B	4-33
21.	C	4-44

Take a short break, then turn to Chapter 5 and continue.

CHAPTER 5

TUBING, PIPE, AND EXTRUSION DISCONTINUITIES

Pipe can be, and has been, manufactured from a variety of materials including concrete, plastic, nonferrous and ferrous. Ferrous simply means pertaining to iron, and it is generally the most widely manufactured pipe. The material selected is dependent on the application. Considerations in design include environment, medium carried, weight, strength, stress handling capability, and other engineering concerns.

The type and significance of discontinuities in pipe also varies with the manufacturing process. We will focus on manufacturing discontinuities. Pipes and tubes fall into two categories: welded and seamless. Welded pipe has a seam along its length, but seamless pipe has no seam or weld.

Welded Tubing and Pipe

Welded tube is made from a narrow, flat piece of steel called skelp. The skelp is heated to a welding temperature and then grasped by a pair of tongs and pulled through a die called a welding bell.

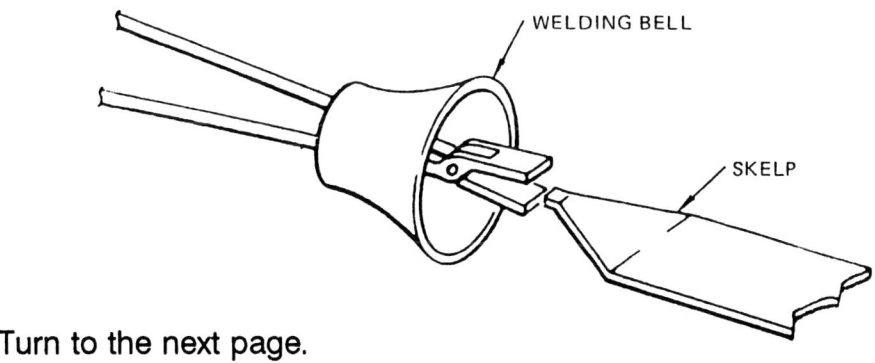

Turn to the next page.

The tongs are then attached to a moving chain that pulls the skelp through the welding bell. Because the skelp is heated to welding temperature, it welds or fuses when the two edges come together in the shape of a round tube.

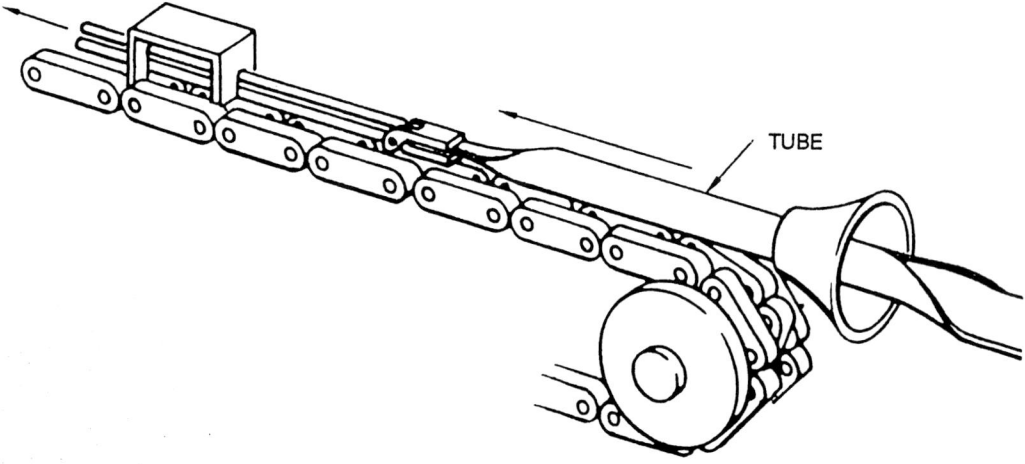

Actually, the material is fused where the two edges come together if the proper temperature is maintained.

Turn to the next page.

From page 5-2   5-3

If the proper temperature is not maintained, there may be intermittent lack of fusion or intermittent cracks. These cracks may appear either on the inside or outside of the tube.

Pipe, which is really thick-walled, large-diameter tube, may be fabricated by other processes. These include hot or cold rolling plate into a cylinder then welding the two mating edges together, forming a longitudinally welded seam.

Which of the following types of discontinuities do you think fit the description of the intermittent crack?

Stringer .... **Page 5-4**
Seam .... **Page 5-7**
Lamination .... **Page 5-9**

Nope. The cracks caused by lack of fusion in the welding of pipe are not called "stringers." A stringer is a discontinuity we learned about when we discussed the working of a billet. It is defined as:

> A NONMETALLIC INCLUSION WHICH IS ELONGATED
> AS THE BILLET IS ROLLED INTO BAR STOCK.

As the skelp is pulled through a welding bell to form tube, the edges should fuse (or weld) when they come together. Intermittent cracks may occur from lack of fusion if the proper temperature is not maintained in the metal.

Turn back to page 5-3 and try again.

From page 5-7   5-5

Here are two typical sections of pipe made by the welding process.

Since the weld runs the length of the pipe, any discontinuity caused by faulty welding would be called a:

hot tear .... Page 5-6
stringer .... Page 5-8
seam .... Page 5-10

From page 5-5

No. It looks as if you have mixed up some definitions. A discontinuity caused by faulty welding in a pipe would not be called a hot tear. A hot tear is a type of discontinuity that occurs in a casting. It is likely to occur at a junction of light and heavy sections of a casting.

A discontinuity in the weld of a welded pipe is something quite different.

Turn back to page 5-5 and make another choice.

Right. Lack of fusion in the welded tube or pipe could cause a seam which might appear either on the inside or outside. Any nick or crack in the skelp or defect in the welding bell die will result in a lap or crack-like discontinuity in the finished tube.

After the tube is formed in the welding bell die, it is put through sizing rollers. These rollers reduce the pipe to its proper size and make it essentially round.

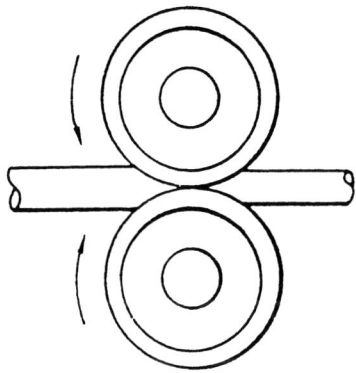

Turn back to page 5-5.

From page 5-5

Wrong choice. It looks as if you have mixed up some definitions. A welding discontinuity in a piece of welded pipe is not a stringer. A stringer is a nonmetallic inclusion in bar stock.

A discontinuity in the weld of a pipe is something quite different.

Turn back to page 5-5 and select another answer.

From page 5-3

That's wrong. The cracks caused by lack of fusion in the welding of tube or pipe are not called "laminations." A lamination is a discontinuity we learned about when we discussed the rolling of a slab to form sheet or plate stock. It is defined as:

A NONMETALLIC INCLUSION WHICH IS FLATTENED AS
A SLAB IS ROLLED INTO PLATE STOCK.

As the skelp is pulled through a welding bell to form tubing, the edges should fuse (or weld) when they come together. Intermittent cracks may occur from lack of fusion at the welding edges if the proper temperature is not maintained in the metal.

This discontinuity would be called a:

stringer .... **Page 5-4**
seam .... **Page 5-7**

That's right. Faulty welding in a pipe would be called a seam. Here is an example of a seam on the inside surface of tubing or pipe. The example has been magnified many times.

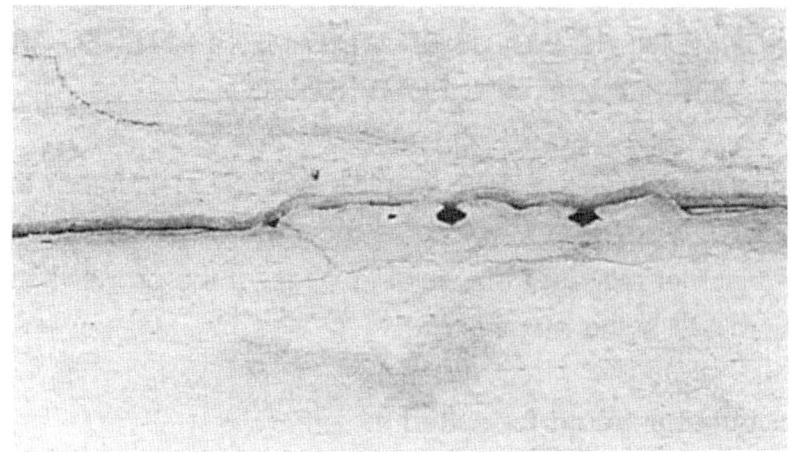

Since welded pipe or tubing is made from sheet or plate material, discontinuities found in them would also be found in the pipe made from these materials. Laminations caused by flattened-out nonmetallic inclusions or porosity would be found in the pipe.

Turn to the next page.

From page 5-10   5-11

Seamless Pipes and Tubes

Seamless pipes and tubes are made from bar stock or billets. In this process, the bar stock is heated to rolling temperature, and a hole is pierced through it lengthwise, forming a seamless tube. The piercing machine has two barrel-like rollers. Between the rollers is a long piercing mandrel with a bullet-shaped nose.

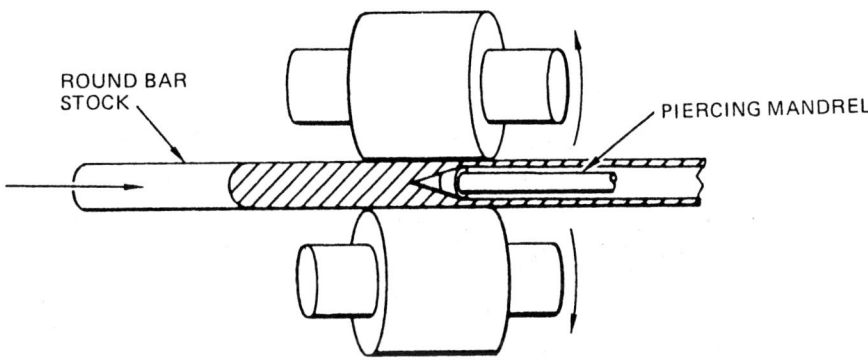

The revolving rollers grasp the white-hot bar stock, spinning it and rolling it over the bullet-shaped piercing mandrel. As the bar moves through the rollers, the piercer goes on through the length of the bar, forming a rough pipe without any seam. As is the case with welded pipe, the seamless pipe is put through sizing rollers to reduce it to proper size and make it round.

Since seamless pipe is made from round bar stock, which of the following discontinuities could appear on the outside surface of the pipe?

Seam . . . . **Page 5-12**
Lamination . . . . **Page 5-15**

Sure. A crack in a billet would be stretched out into a seam in round bar stock. Since seamless pipe or tubing is made from bar stock, the pipe could certainly have a seam.

Seamless pipe can have discontinuities on the inside of the pipe caused by the piercing mandrel. As the white-hot bar stock is spun and rolled over the piercing mandrel, some of the metal pieces will occasionally adhere to the mandrel. The metal buildup on the mandrel may be torn from the mandrel and fused back into the pipe, gouging a rough depression in the process. The pieces of metal fused into the pipe in this manner are called SLUGS.

Turn to the next page.

Here is another example of a slug found in a liquid oxygen line used for fueling a space-launch vehicle.

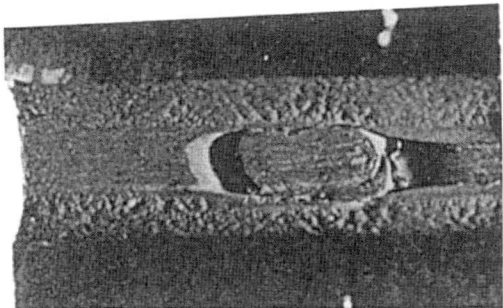

In some cases, pipe or tubing is put through sizing rollers that have a sizing mandrel between the rollers as shown here.

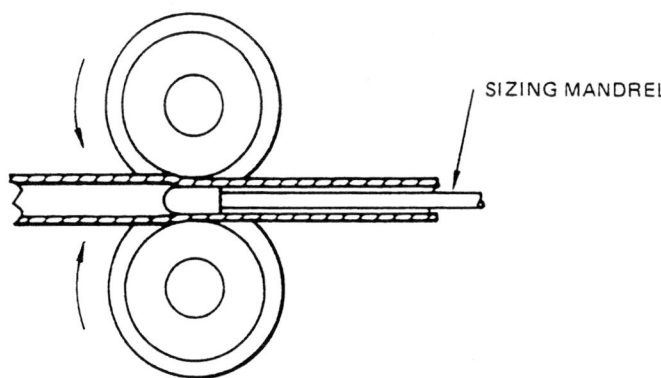

Use of the sizing rollers with the mandrel in between assures that both the inside and outside diameters of the pipe or tubing are round. Under some conditions, friction between the mandrel and the inside surface of the pipe causes gouging of the inside surface of the pipe.

Turn to the next page.

From page 5-13

5-14

Here is an example of gouging on the inside surface of a pipe.

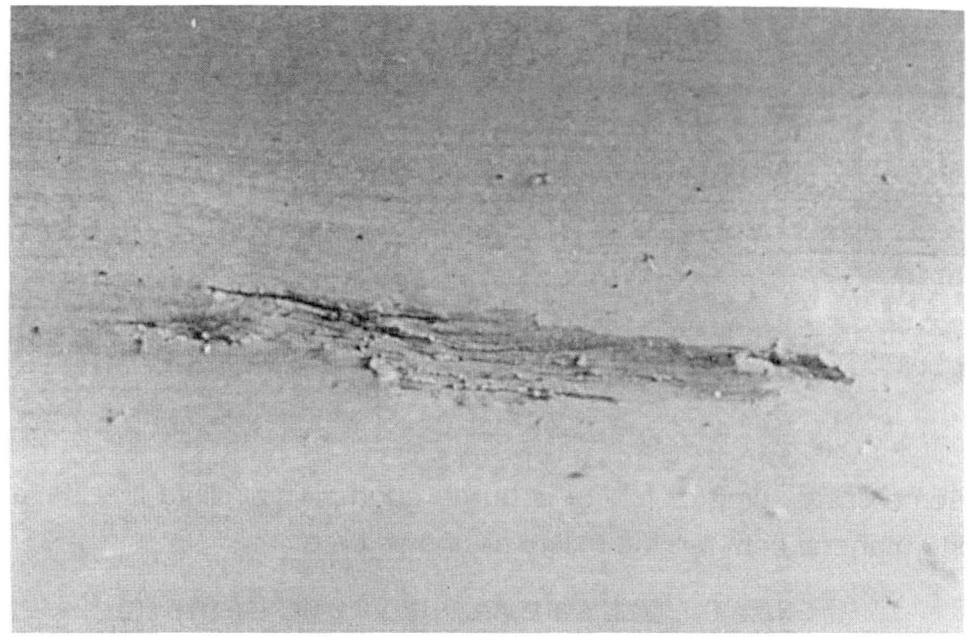

Here is a discontinuity on the inside surface of a piece of pierced pipe.

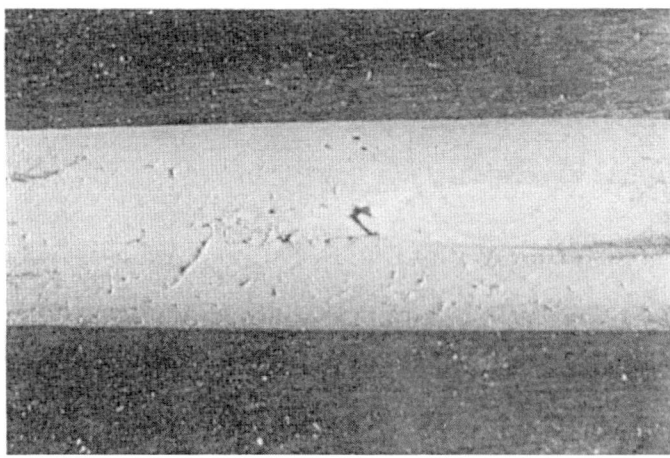

What is the name of this discontinuity?

Gouging . . . . **Page 5-16**
Slug . . . . **Page 5-19**

You believe that seamless pipe could have laminations. That choice was wrong. A lamination is found in sheet or plate stock. It is a nonmetallic inclusion that is flattened out when a slab is rolled into *plate stock*.

Seamless pipe is not made from plate stock. It is made from round bar stock and is less likely to have laminations. But round bar stock can have a discontinuity that would run longitudinally along the outside surface of the pipe after processing.

Turn back to page 5-12 and continue.

From page 5-14

You believe that the discontinuity shown is called gouging. That answer is partially right. Gouging is caused by friction of the mandrel and the inside of the pipe; however, in the case shown, a slug has been deposited in the pipe wall as the result of severe metal buildup on the mandrel. A simple gouging action would look more like the one pictured below. The slug would not be present.

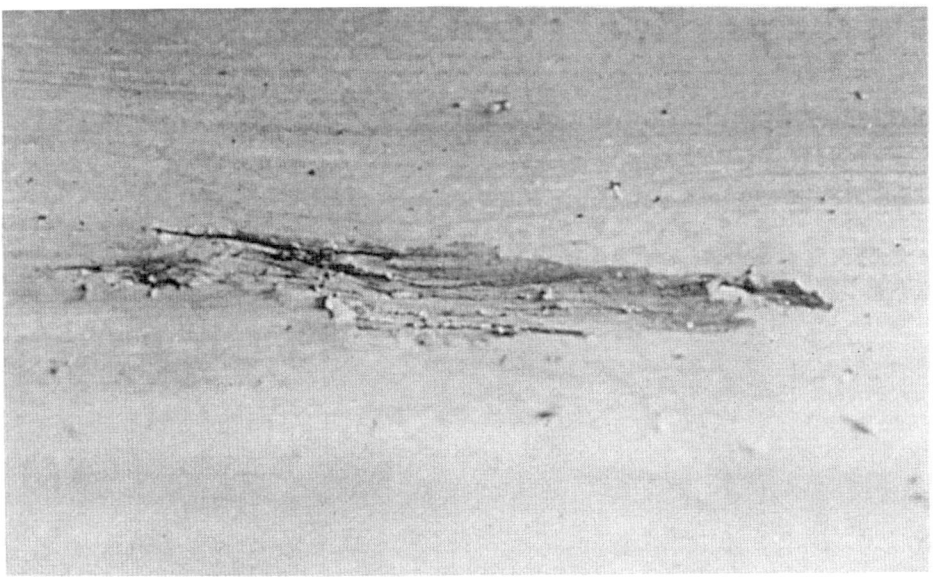

Turn back to page 5-14 and review the photos.

From page 5-19

Welding discontinuities can also be found in welds in welded pipe but will be discussed in a later chapter. In summary, WELDED PIPE OR TUBING can have the following discontinuities:

- SEAMS from lack of fusion at the weld.

- LAMINATIONS which may be in the plate stock or skelp from which the pipe is made.

SEAMLESS PIPE OR TUBING can have the following discontinuities:

- SLUGS caused by metal buildup on the piercing mandrel with subsequent fusing of pieces of the metal to the inner wall of the pipe.

- SEAMS OR STRINGERS found in the bar stock from which the pipe is made.

- GOUGING caused by friction between the sizing mandrel and the inside surface of the pipe or tube.

Turn to the next page.

Extrusion Discontinuities

The forming of parts by forcing metal through a die is known as extrusion. Depending upon the manufacturing requirements, the process may use heated metal (hot extrusion) or metal at ambient temperature (cold extrusion). In either instance, as the metal is forced into or through the die, it takes the cross-sectional shape of the opening in the die.

The extrusion process is adaptable to forming parts from steel, copper, aluminum, and other metals and alloys. The two pieces below are examples of aluminum extrusions.

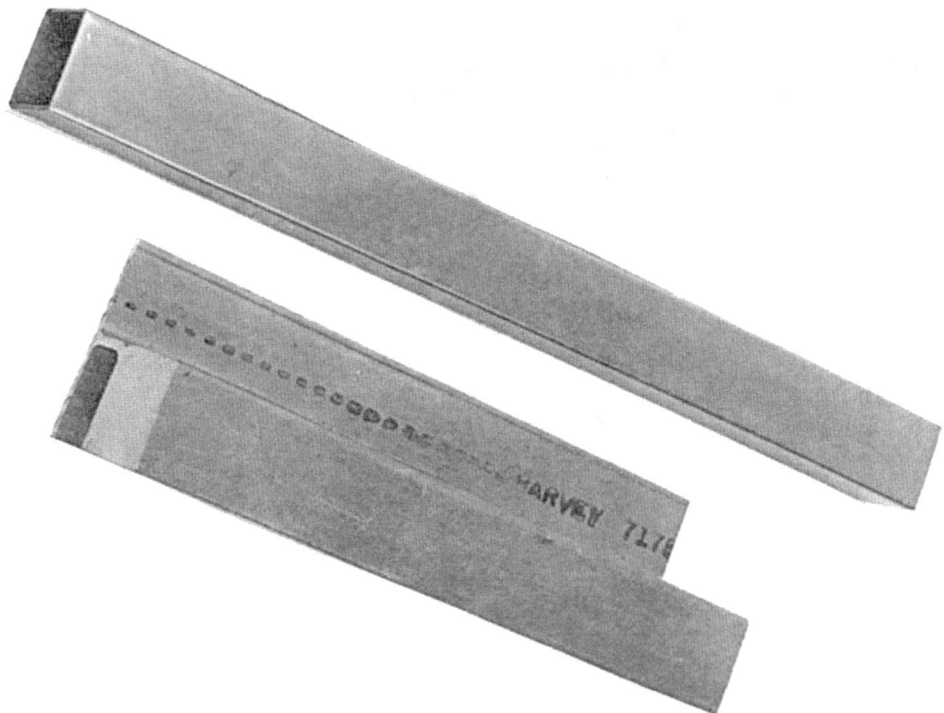

Turn ahead to page 5-20.

From page 5-14

Absolutely. That discontinuity was a slug in a piece of pierced pipe, and it was caused by metal buildup on the piercing mandrel. Here is another example of slug.

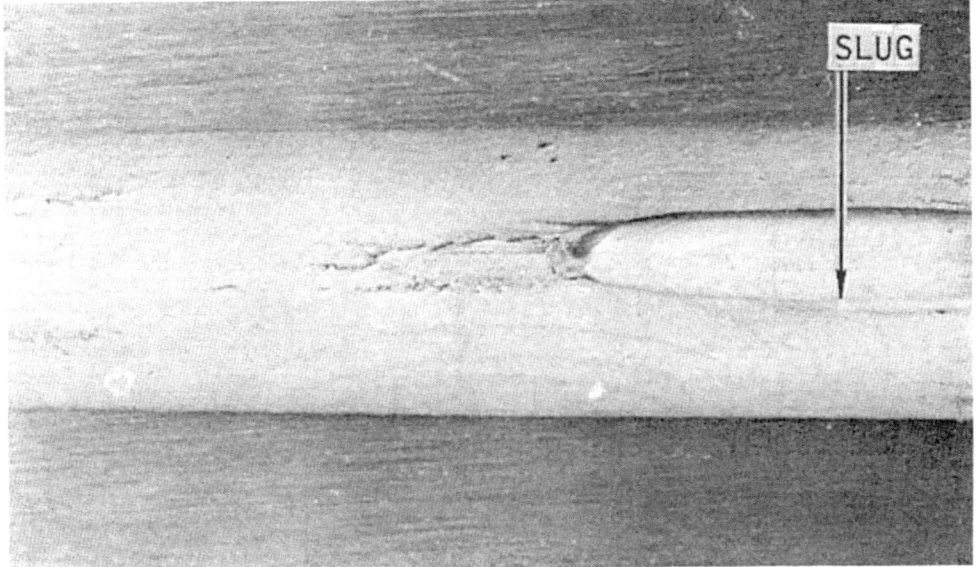

Turn back to page 5-17.

From page 5-18

Let's look at a soft steel part formed by using a punch to force the metal through the die.

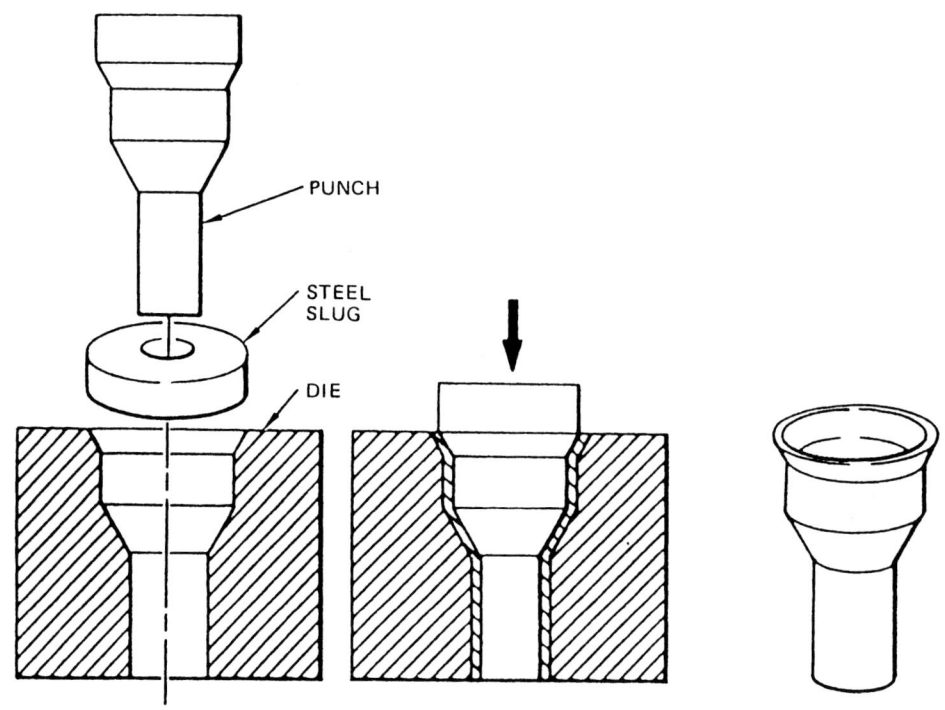

Assuming that the steel slug has been roughed out from bar stock, what types of discontinuities would possibly be in the formed part?

Laminations and hot tears .... Page 5-21
Seams and porosity .... Page 5-22

From page 5-20

No. Laminations and hot tears are not likely to be found in an extruded part formed from steel bar stock.

Laminations occur in processing metal by rolling it into sheet stock. They are nonmetallic inclusions that are spread out or sandwiched into sheet stock as it is formed.

Hot tears are normally associated with casting. They occur from the differential cooling of light and heavy sections of a casting.

The point here is that formed parts can have the same discontinuities as the metal from which they are formed. If the original bar stock had contained a crack or porosity, the same discontinuities would show up in the formed part. The part could have seams and porosity.

Turn to the next page.

Sure. The extruded part was formed from bar stock. If the bar stock had contained cracks (seams) or porosity, the formed part would have the same discontinuities.

The extrusion process itself may cause discontinuities other than those associated with the stock from which a part was formed. If the metal does not "flow" through the die properly, the finished part will contain discontinuities from the extrusion process. In hot extrusions, the metal must be at a temperature that it can be forced into the desired shape. In cold extrusions, the metal must "cold-flow" to form the proper shape.

Nondestructive testing requirements for extruded parts will be determined on the basis of the metal used, the complexity of the part, and other factors inherent in the extrusion process.

Turn to the next page for a quick review of tubing, pipe, and extrusion discontinuities.

CHAPTER REVIEW

____ 1. Pipes and tubes are classified into two basic categories which are:

   A. welded and seamless.
   B. welded and cast.
   C. seamless and forged.
   D. cast and forged.

____ 2. In welded tubing, a discontinuity caused by lack of fusion in the weld is called a:

   A. weld spot.
   B. stringer.
   C. seam.
   D. de-fusion.

____ 3. A seam in welded tubing may occur on the:

   A. inside.
   B. outside.
   C. inside or outside.
   D. tube away from the weld.

From page 5-23

_____ 4. Seamless tubing is made from round bar stock. A crack in the stock may appear as a _____ on the outside of the tubing.

A. stringer
B. seam
C. gouge
D. slug

_____ 5. The piercing mandrel can cause two types of discontinuities which are:

A. gouges and stringers.
B. gouges and seams.
C. seams and stringers.
D. gouges and slugs.

_____ 6. _____ are caused by metal that builds up on the piercing mandrel and then breaks off and fuses to the pipe wall.

A. Gouges
B. Slugs
C. Seams
D. Stringers

7. Extruded parts may contain any of the discontinuities that were in the _____ from which they were formed.

   A. stock
   B. mandrel
   C. weld
   D. seams

Turn to the next page for answers to these review questions.

From page 5-25

ANSWERS TO REVIEW QUESTIONS FOR CHAPTER 5

Question & Answer	Reference Page(s)
1. A	5-1
2. C	5-3, 5-7
3. C	5-7
4. B	5-12
5. D	5-16
6. B	5-12
7. A	5-22

Turn to Chapter 6 for a discussion of processing discontinuities.

CHAPTER 6

PROCESSING DISCONTINUITIES

In this chapter we will look at discontinuities associated with the following processes: grinding, heat treating, explosive forming and fatigue cracking. Let's begin by looking at cracks related to grinding.

Grinding cracks can be caused by stresses built up from excess heat created between the grinding wheel and the metal.

GRINDING CRACKS

During the grinding of metal, heat occurs at the point of contact between the grinding wheel and the metal. The wheel grinds, heats, and expands the metal directly beneath it. The wheel passes, and the small area that has just been ground cools and shrinks.

According to the illustration, grinding cracks:

will be found across the grain .... Page 6-2
occur at a right angle (crosswise) to grinding wheel rotation .... Page 6-5

From page 6-1

Cracks will not necessarily be found across (or with) the direction of the grain because grain direction has nothing to do with crack direction. The picture on page 6-1 indicated that the cracks occurred at right angles or crosswise to wheel rotation. In this instance, the cracks also happened to go across the grain.

Turn ahead to page 6-5 and continue.

From page 6-5

No, grinding cracks do not always go with the grain in hard metals. Remember that crack direction has nothing to do with grain direction.

Turn ahead to page 6-6.

From page 6-6

Heat Treating

Basically, heat treating is the process of hardening or softening metals by controlled heating and cooling. It is a process by which desired mechanical properties can be introduced into metals. Heat treating, for instance, can be used to give machinability to metal to be machined.

Heat and the processing of steel are closely related. Iron ore is smelted in blast furnaces, castings are poured from molten steel, and forging is done on preheated metal. Throughout the steel-making and manufacturing processes, steel is heated and cooled, heated and worked, and heated and reworked. These steps, which produce desired qualities in the metal, can also cause undesirable side effects.

These undesirable aspects of metal processing are called:

defects .... **Page 6-7**
discontinuities .... **Page 6-8**

Right. Grinding cracks occur at a right angle (crosswise) to the rotation of the grinding wheel. These cracks have no relation to the grain direction of metal.

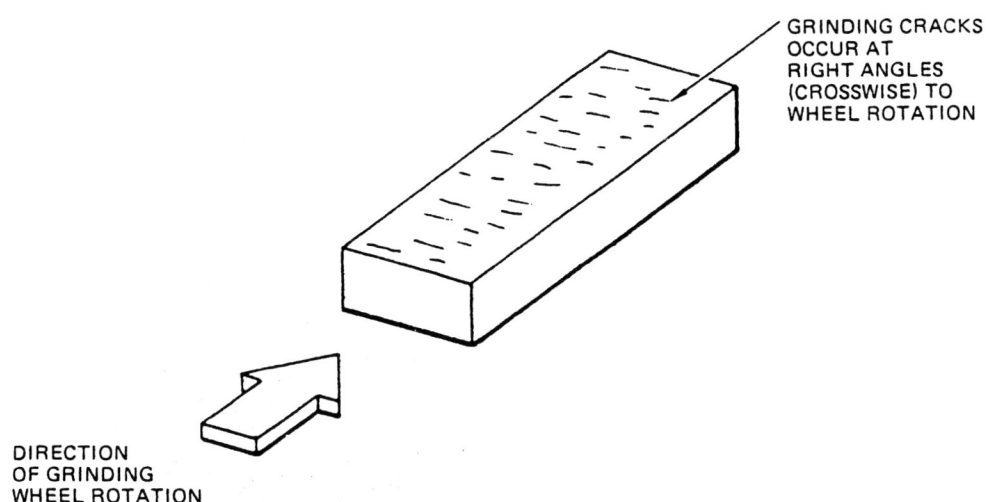

When grinding cracks occur, they will form at right angles to the direction of wheel rotation. When too much heat is allowed to build up on the surface of the metal being worked, stresses occur. These stresses relieve themselves in the form of cracks.

Very hard heat treated or plated metals are the metals most apt to crack from grinding.

Which is true?

Grinding cracks always go with the grain in hard metals . . . . Page 6-3
Grinding cracks might or might not occur across the grain in hard metals . . . . Page 6-6

Good. Grinding cracks *may or may not* occur across the grain in hard metals because grain direction is not a factor governing crack direction. Grinding wheel rotation provided the clue to crack direction, since grinding cracks (when they occur) occur *crosswise to wheel rotation*.

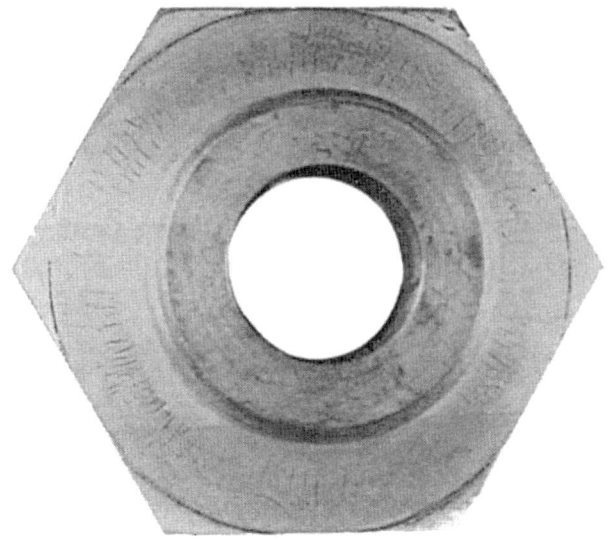

The above illustration shows grinding cracks. Poor grinding technique caused the part to become overheated, and the cracks resulted. Any cracks are bad, but the worse the grinding technique, the worse the grinding cracks become. If extreme heat is generated, originally crosswise grinding cracks can extend in many directions, forming a lattice-work or checkerboard pattern of cracks.

Turn back to page 6-4.

From page 6-4

You chose the term "defects" to describe the undesirable aspects of metal processing. This term is not correct. Although the undesirable aspects of metal processing might later be labeled *defects*, depending on several factors, they are first discontinuities, because discontinuities are breaks in metal structure.

Turn to the next page.

From page 6-4   6-8

Certainly. Discontinuities - the undesirable aspects of metal processing - are the topics being discussed, and discovering discontinuities is the purpose of nondestructive testing.

During the heat treating process stresses are built up which, if not relieved by proper control, will find relief through cracking. A corner, a tool nick, or a burr provide a starting point for heat treated cracks. These sharp corners act as stress-concentration points in the metal.

Unlike seams, that occur only in the direction of grain flow, or grinding cracks, that have their beginning crosswise to wheel rotation, heat treatment cracks have no specific direction.

Heat treatment cracks:

might follow the grain or cross the grain . . . . **Page 6-10**
would probably follow the grain . . . . **Page 6-12**

From page 6-13

Right. Metal has the property of shrinking as it cools, which creates discontinuities. Metal occupies more space when hot or molten than when solid or cool. In heat treating, the light sections cool faster than the heavy sections that they join, and cracks can appear at these light and heavy junctions.

Heat treating, then, can cause discontinuities just as can the other metalworking and forming processes. During inspection of heat treated parts, the first areas of concern will be:

- Any sharp area, such as corners, ridges, etc.

- Junctions of light and heavy sections.

Turn ahead to page 6-15.

From page 6-8   6-10

Yes, the answer "Heat treatment cracks might follow the grain or cross the grain" is merely restating that heat treatment cracks have no specific direction.

The photo below shows a part that has cracked from heat treatment. Notice that the crack cuts across the grain. Since heat treatment cracks have no specific direction, the crack might just as well have followed the grain.

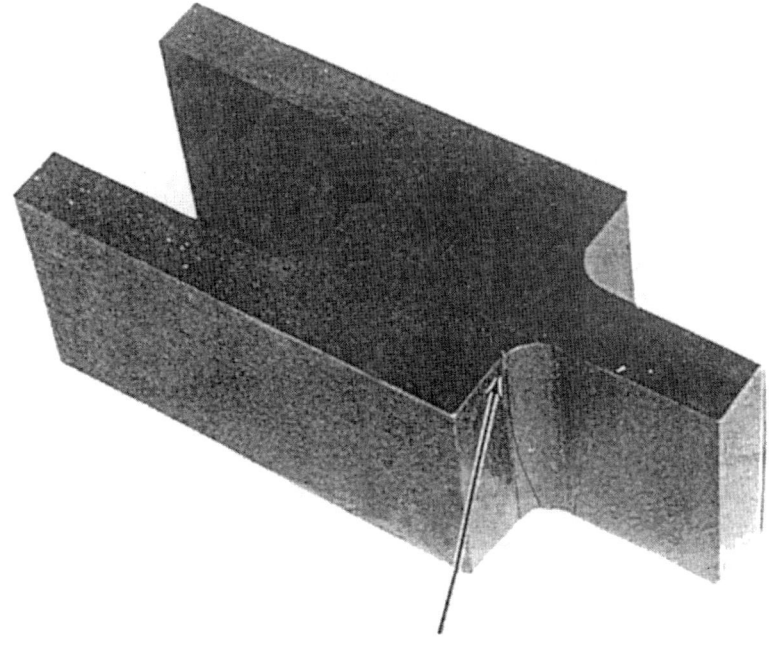

Notice carefully the spot pointed out by the arrow. It indicates that the crack started from a:

small ridge which had not been machined away . . . . Page 6-13
smooth, curved part of the metal . . . . Page 6-14

From page 6-13

No, metal does not expand when it begins to cool or harden. Regardless of the metal-forming processes, metal shrinks (contracts) as it hardens. This sentence is just another way of stating that metal occupies more space when hot or molten than when solid or cool.

Turn back to page 6-9.

From page 6-8

No, the answer that heat treatment cracks would probably follow the grain is incorrect. Remember that they occur in no specific direction.

Turn back to page 6-10.

Very good. You spotted the small, unmachined ridge. The part was improperly heat treated, and the ridge acted as a stress-concentration point for the crack. Although the ridge was very slight, any sharp area, whether small or large, can act as a stress-concentration point during heat treating of metal.

Unequal cooling between light and heavy sections of a part being heat treated can also result in cracking.

As you probably remember, this is a result of the property that metal has of:

occupying more space when hot or molten than when solid or cool .... **Page 6-9**
expanding when cold .... **Page 6-11**

From page 6-10

This was a difficult one to see, and you are certainly not to be chastised for not spotting the very small ridge in the part that acted as a stress-concentration point. Nevertheless, there is a ridge in the curved area.

Turn back to page 6-10, take another look and, even though you might still not see it, turn to the page that indicates that the ridge is there.

Explosive Forming

In another process, explosive forming, an explosive charge is used to generate a pressure which "forms" a metal part in the die. The charge is supplied by a variety of explosives ranging from shotgun shells to high explosives such as dynamite.

The forming may be done either in "open air" or submerged in a liquid, such as water. In the latter instance, the water supplies a means of transmitting the shock wave of the explosion to the part to be formed. The simplified illustration below depicts explosive forming in water.

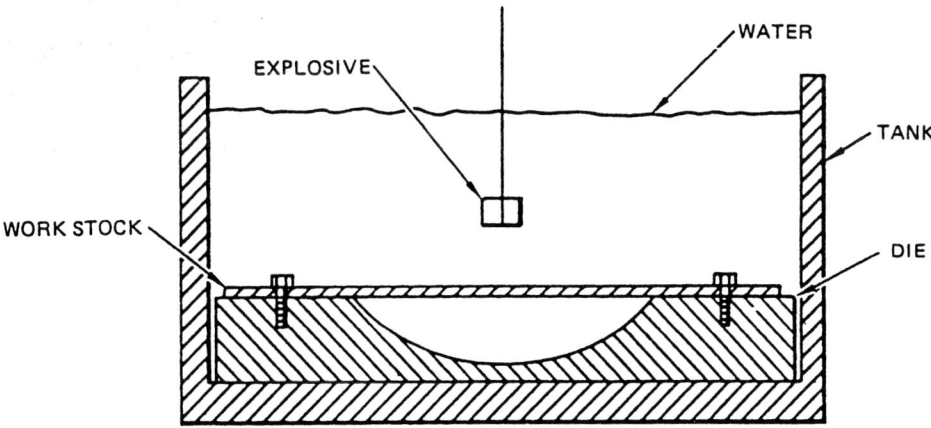

The finished part from a cupping die such as the one above (after the flange is trimmed) would look something like this:

Turn to the next page.

From page 6-15

In the illustration on the preceding page, the water transmits the shock wave from the explosive to the stock, forcing it into the die to form the desired shape.

The other method of explosive forming (open air) is accomplished without the use of water. It is done by placing charges at strategic locations within a metal blank (such as a tube) that is confined inside a die. The explosion forces the blank to conform to the shape of the die. The photograph below shows one-half of a die with a metal blank inserted for forming.

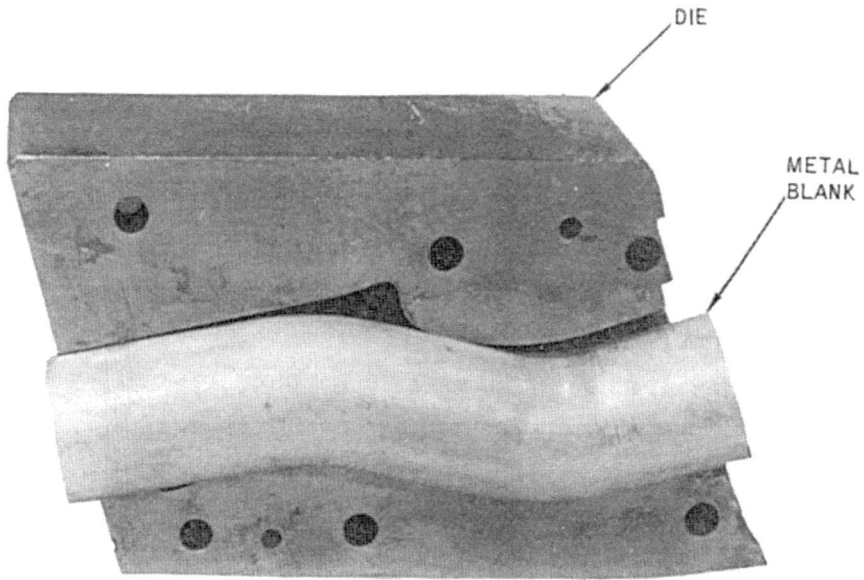

Turn to the next page.

From page 6-16

Now let's take a look at the finished product with the die opened following the explosive forming of the part.

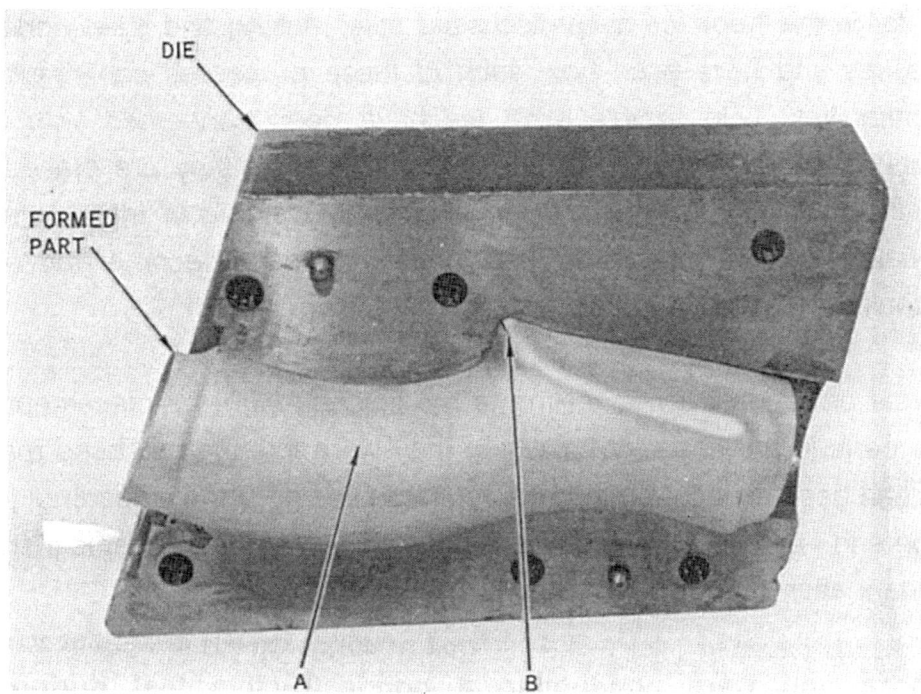

You can see that the explosion caused the blank to conform to the shape of the die. Study the photo carefully.

Do you think a discontinuity would most likely occur at point A or point B?

A . . . . **Page 6-20**
B . . . . **Page 6-22**

From page 6-22   6-18

Fatigue Cracks

Thus far in this book we have discussed steel-making and steel-working processes and have seen how each of these processes can result in discontinuities. Up to this point we have been concerned with the condition and quality of parts and materials before they are placed in service. Now let's look at a discontinuity that can occur in parts and materials that have passed all inspections and have become part of a functioning industrial product.

The role of nondestructive testing is not finished with the placement of these parts in the capacity for which they were intended. These metal parts can become fatigued through repeated use and/or overloading, just as you and I can. The role of nondestructive testing, therefore, is just as important after the parts are placed in service as before.

Suppose you were conducting a nondestructive test during a maintenance inspection on an airplane. Would you consider the landing gear an area of high stress requiring inspection?

Yes . . . . Page 6-21
No . . . . Page 6-23

Fatigue cracks on highly-stressed surfaces (such as the landing gear beam) often start from discontinuities. Nicks, grinding cracks, forging laps, even poorly-finished surfaces are all examples of discontinuities that might result in fatigue cracks.

What is your opinion of porosity or nonmetallic inclusions within a highly-stressed metal part? Could they possibly cause fatigue cracks?

No .... Page 6-25
Yes .... Page 6-26

From page 6-17   6-20

You selected A as the point at which a discontinuity is most likely to occur. Your selection is wrong. Here are the die and the formed part.

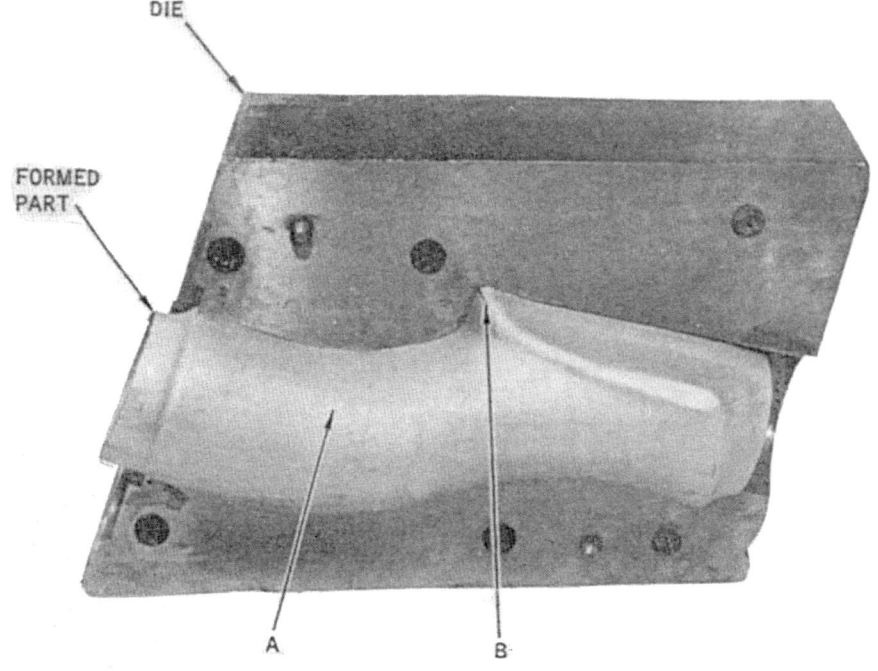

Take another look. Point A is located in an area where the metal was not greatly affected by the explosion. The metal did not have to be deformed much at that point to conform to the shape of the die.

By comparison, at point B the metal had to be sharply changed in shape to conform to the shape of the die. In fact, if you take a close look at the formed part, you can detect a crack or tear in the part at point B.

Turn ahead to page 6-22.

Yes, if conducting a nondestructive test on an airplane during a maintenance inspection, you would most certainly inspect the landing gear -- an area of high stress! Each time the airplane lands, the landing gear is subject to high stress. The harder the landing or the heavier the airplane, the greater the stress.

Some time ago, a heavily-loaded helicopter collapsed while parked. It was discovered that the main support beam on the landing gear had sustained a tool nick. The nick (a discontinuity) created an area of stress concentration from which a fatigue crack had formed. The crack progressed slowly through the beam until it failed, without warning, and the landing gear collapsed.

Turn back to page 6-19.

Correct. The part shown is most likely to have a discontinuity at point B. In fact, if you take a close look, there is a discontinuity at point B.

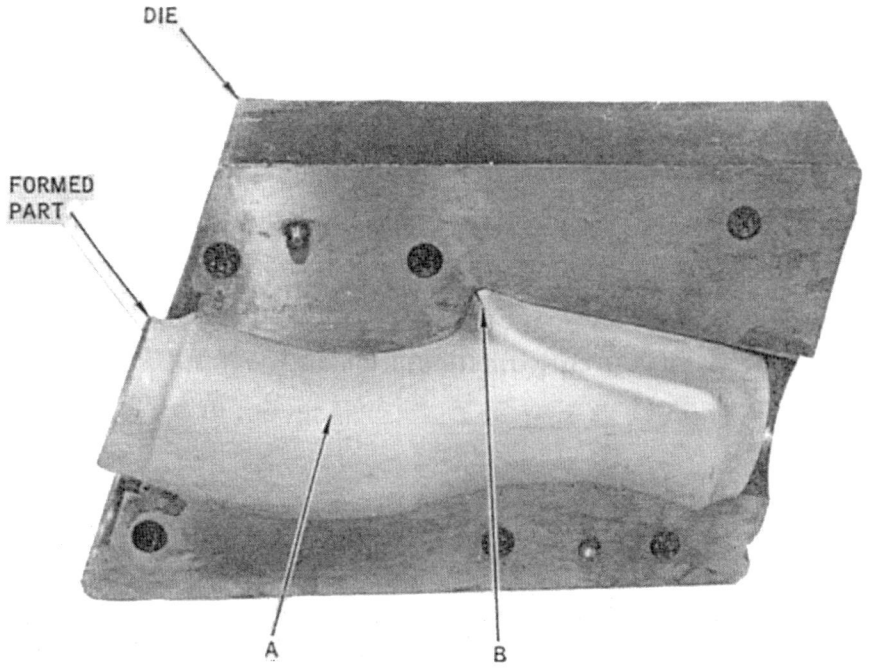

The discontinuity is a crack or tear caused when the explosive force overstressed the material while shaping it to the contour of the die. A discontinuity is most likely to develop at a point in the die where the most extreme deforming of the stock occurs - or where the die changes contour abruptly.

Explosively-formed parts can have discontinuities caused by stress in the forming, improper die junctions, or ones that existed in the metal stock or blank from which the part is formed.

Our next subject deals with discontinuities in parts that passed former nondestructive testing inspections and that have been in use.

Turn back to page 6-18.

From page 6-18

You said that an airplane landing gear is not an area of high stress, but it is. The landing gear of an airplane must support the weight of the entire airplane while the plane is on the ground. Additional stress is placed on the landing gear during hard landings.

Turn back to page 6-21.

From page 6-26

6-24

Highly-stressed surfaces are subjected to movements in several directions or heavy loading. A wire bent rapidly back and forth is an example of metal stress.

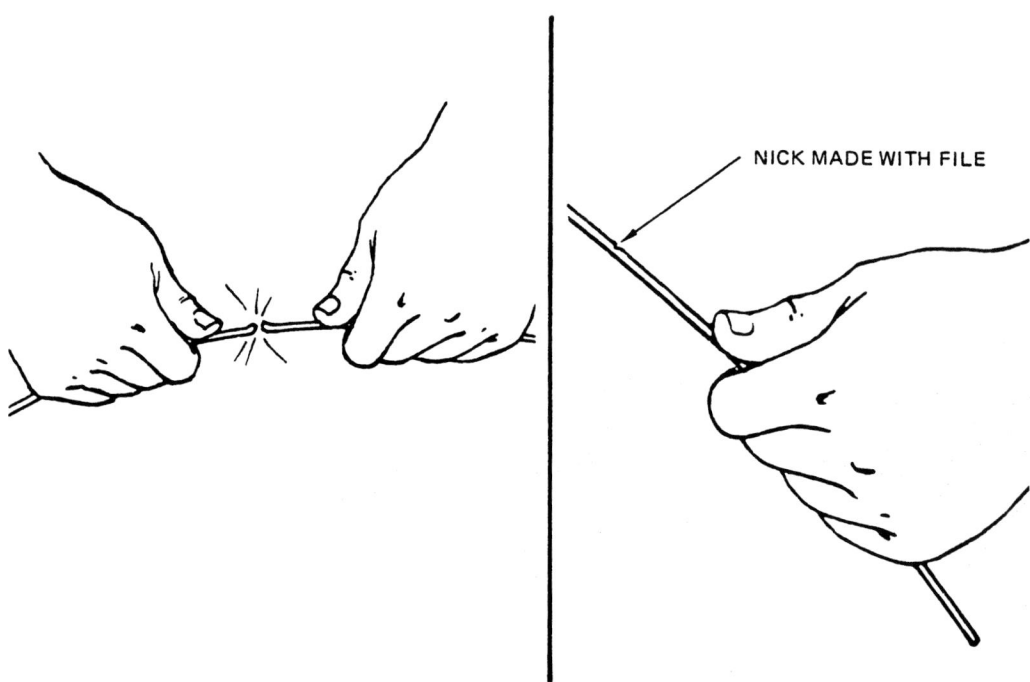

The wire in the left picture broke after 12 bends. An as-yet-unbroken wire from the same roll (right picture) has been nicked slightly with a file.

The nicked wire will:

require 12 bends before breaking .... Page 6-27
break before being bent 12 times .... Page 6-28

From page 6-19

Your answer that "no, nonmetallic inclusions or porosity could not possibly cause fatigue cracks" is wrong. Even though most fatigue cracks occur from surface discontinuities, subsurface discontinuities also create stress-concentration points and can cause fatigue cracks.

Turn to the next page.

From page 6-19

Excellent. Porosity or nonmetallic inclusions within a highly-stressed metal part could certainly result in fatigue cracks. These discontinuities serve as stress-concentration points (as starting points) for fatigue cracks.

Although some fatigue cracks might be subsurface, such as those having their beginning from porosity or nonmetallic inclusions, most fatigue cracks are open to the surface. Why? Fatigue cracks usually start from stress-concentration points, which themselves are open to the surface.

Turn back to page 6-24.

From page 6-24

6-27

No. The nicked wire would *not* require the same number of bends (12) to break as the unnicked wire required.

Stress will concentrate at the nick in the wire, and the nicked wire will break sooner than did the unnicked wire, which has no stress-concentration point.

Turn to the next page.

Right. The unnicked wire broke after 12 bends. The other wire was given a stress-concentration point and will certainly break before being bent 12 times.

Fatigue cracks in functioning parts occur crosswise to the direction of stress movement. The fatigue crack shown in the drive shaft below illustrates this fact. The stress on this shaft would have been clockwise - the direction of its rotation. The fatigue crack occurred across the direction of stress movement.

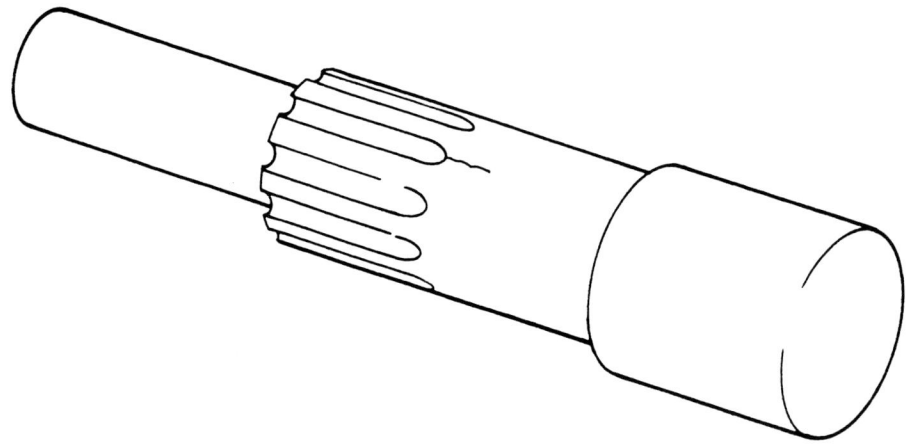

Fatigue cracks are possible only after a part has gone into service. If undetected, they will eventually result in failure of the part and possibly of the entire product to which it belongs. But they can be found through nondestructive testing, just as can those discontinuities found in parts and materials during the steel-making and steel-working stages.

For a review of Chapter 6, turn to page 6-29.

CHAPTER REVIEW

1. Grinding cracks are one of the easiest types of discontinuities to recognize. They are caused during grinding when:

    A. too little pressure is used.
    B. too much heat is generated.
    C. too little heat is generated.
    D. an explosion occurs.

2. Grinding cracks occur _____ the direction of grinding.

    A. with
    B. across
    C. independent of
    D. radially to

3. Since grinding cracks occur in a direction that is dependent on the wheel rotation, their growth:

    A. is similar to seams.
    B. is highly dependent on grain direction.
    C. is depend on grinding wheel material.
    D. has nothing to do with grain direction.

From page 6-29   6-30

_____ 4. In severe cases of grinding cracking, the cracks may:

   A. grow in a line parallel to the grinder rotation.
   B. be entirely subsurface.
   C. form a lattice-work or checkerboard pattern.
   D. form a "star" pattern.

_____ 5. Discontinuities can be prevented during grinding if too _____ is NOT generated.

   A. much heat
   B. much grinding wheel rotation
   C. little heat
   D. little wheel rotation

_____ 6. If grinding cracks do occur, they have their beginning in a:

   A. direction with grinding wheel rotation.
   B. direction crosswise to the grinding wheel rotation.
   C. direction independent of grinding wheel rotation.
   D. checkerboard pattern.

___ 7. A process which can cause discontinuities while heating metal to cause desirable qualities is called:

   A. checkerboarding.
   B. fatigue processing.
   C. heat treatment.
   D. nicking.

___ 8. The heating of a component or part will cause stresses present during the shaping processes to be released. Areas such as corners, nicks, and burrs are subject to this concern and are referred to as:

   A. stress concentrations.
   B. heat treatments.
   C. fatigue points.
   D. lattice flaws.

___ 9. Heat treatment cracks grow:

   A. in any direction.
   B. with the grains.
   C. crosswise to the grains.
   D. in checkerboard patterns.

___ 10. Thickness changes in a part may also cause cracks. Why?

   A. These changes cause slow cooling rates.
   B. Such parts usually crack anyway.
   C. This results from poor mold design.
   D. These changes result in uneven cooling rates.

From page 6-31

11. Heat treating is used to produce certain desirable qualities in a finished product. Nondestructive evaluation does NOT end here because:

   A. service-related failures may occur.
   B. all heat treated parts will eventually fail.
   C. grain growth must be monitored.
   D. NDE technicians always need jobs.

12. The term which best describes the condition of a part after multiple loadings and high stress levels is:

   A. stress.
   B. concentration.
   C. die.
   D. fatigue.

13. Fatigue cracks occur across the direction of stress. They originate in stress concentration areas. If a fatigue crack started at a tool nick, it would be a(an) _____ and _____ crack.

   A. closed, subsurface
   B. open, surface
   C. open, subsurface
   D. closed, surface

14. If a fatigue crack originated from a subsurface discontinuity such as porosity, it would most likely be a:

    A. surface crack.
    B. blow hole.
    C. subsurface crack.
    D. seam.

15. The forming of parts by the use of an explosive charge is called:

    A. cracking.
    B. heat treating.
    C. explosive grinding.
    D. explosive forming.

16. When forming parts with the use of explosives, the part is forced to fit the shape of a:

    A. grain.
    B. die.
    C. heat.
    D. fatigue.

From page 6-33

_____ 17. A discontinuity is most likely to occur when explosives are used where the part has to conform to abrupt changes. These discontinuities are called:

A. tears.
B. pops.
C. rips.
D. seams.

Turn to the next page for the answers to these review questions.

ANSWERS TO REVIEW QUESTIONS FOR CHAPTER 6

Question & Answer		Reference Page(s)
1.	B	6-1
2.	B	6-5
3.	D	6-6
4.	C	6-6
5.	A	6-6
6.	B	6-5
7.	C	6-4
8.	A	6-8
9.	A	6-8
10.	D	6-9
11.	A	6-18
12.	D	6-18
13.	B	6-26
14.	C	6-26
15.	D	6-15
16.	B	6-22
17.	A	6-22

Turn to the next page and begin Chapter 7.

CHAPTER 7

WELDING DISCONTINUITIES

Welding is a process whereby materials, mostly metals, are joined by heat and sometimes pressure. Welding examination is routine in many industries today. Complex geometries of welding joints, along with exotic materials, make our NDE job an important and challenging one. A flawless, properly-made weld is essential in components, assemblies, structures, and designs. A faulty weld can mean the failure of a part, failure of a system, or failure of an entire program.

Many types of welding processes are used in manufacturing. We will investigate two of the three arc welding processes listed below:

- Gas tungsten arc (GTAW)
- Shielded metal arc (SMAW)
- Gas metal arc (GMAW)

Additionally, each type of weldment has its own techniques and variations. It is not our intent to give you a complete cram course in welding techniques; such an effort would take another few hundred pages. However, you should be familiar with the more common discontinuities found in welds. Each welding process possesses its own potential discontinuities, so it is important that we understand how they work in order to make the appropriate NDE method selection. We will touch briefly on welding process discontinuities in this chapter.

Turn to the next page.

From page 7-1 7-2

Welding Terms

The illustration below shows the typical welding terms that will be used in this section. You may want to mark this page for reference.

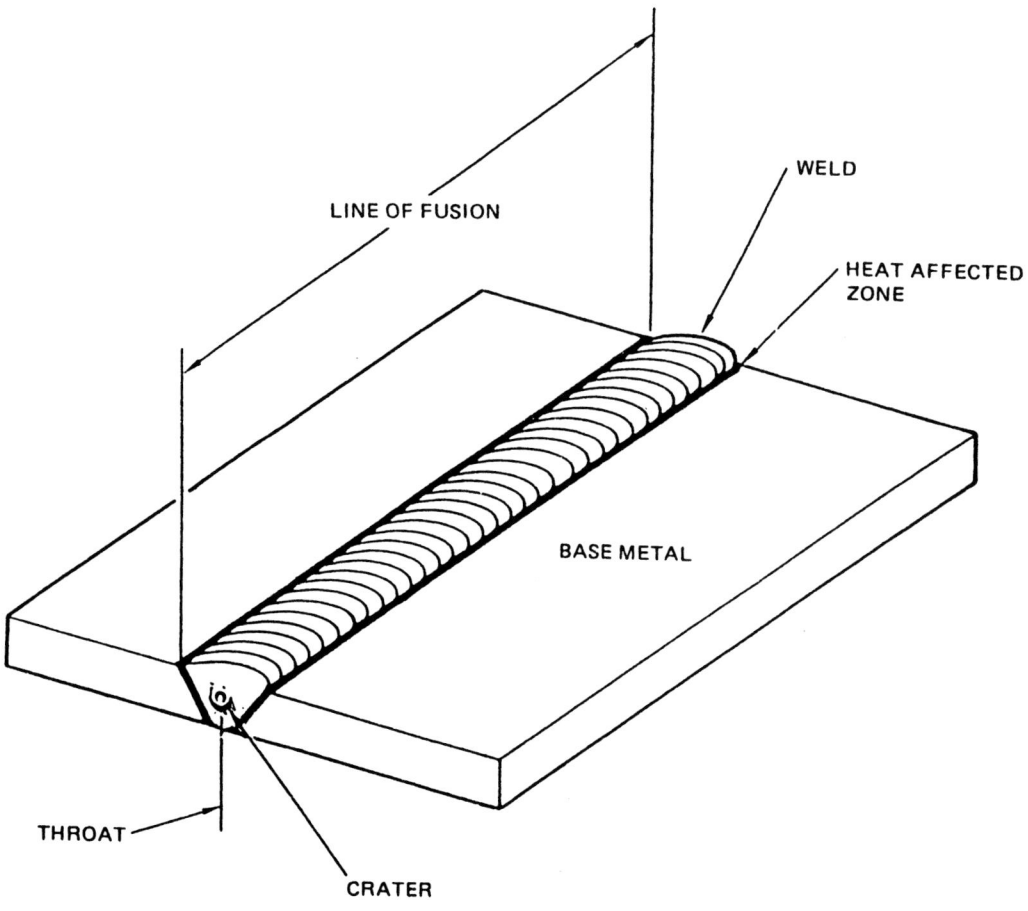

Turn to the next page.

Crater Cracks

Crater cracks are caused at the weld-bead crater by improper use of the heat source either when a weld is started or stopped. The start of a weld and the end of a weld, then, would be obvious places to look for crater cracks. But suppose that a weld was stopped or broken off temporarily. A crater crack could occur at this temporary stop. The illustrations below show welds in different stages of completion. In each of the welds a crater crack has occurred. You can see that these cracks can take different shapes.

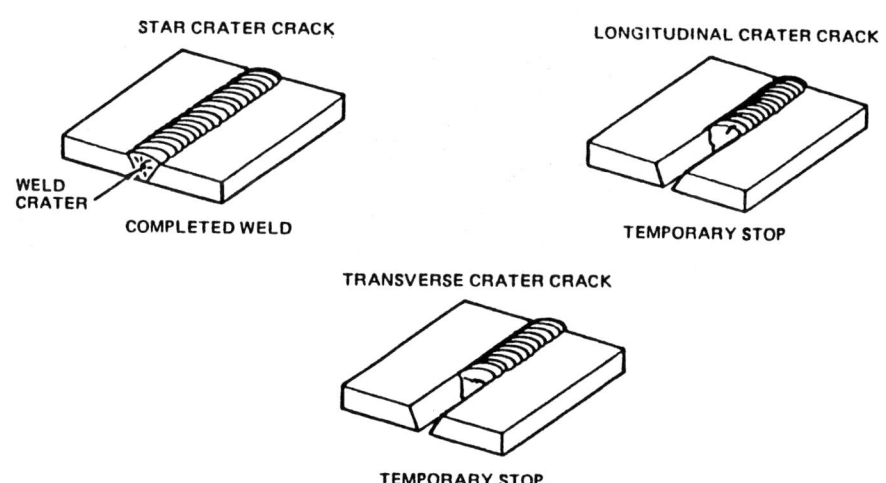

The two welds on the right above have not been completed. When the welder takes up his torch or arc to complete the welds, he must take care to fuse the discontinuity together. If he does not completely fuse the metal, a crater crack will remain. **Where *might* a crater crack in a completed weld be found?**

At the end of the weld . . . . Page 7-4
At the beginning of the weld . . . . Page 7-7
Somewhere between the beginning and end of the weld . Page 7-9
At any one, or possibly all, of the above locations . . . . . Page 7-12

From page 7-3 7-4

The answer you have chosen - a crater crack in a completed weld might be found at the end of the weld - is only *partially* correct. If the heat source were improperly used at the end of the weld, a crater crack might have occurred there.

The same reasoning applies to the *start* of the weld and somewhere between the beginning and end of the weld (the weld might have been stopped or broken off temporarily). The occurrence of a crater crack depends on whether the heat source was improperly used at the sites.

Turn ahead to page 7-12 and continue with the discussion.

From page 7-12

You are evidently not recalling the crater crack drawings which have been shown, because crater cracks are not all shaped alike and they can and do take different directions.

Turn ahead to page 7-8 for a discussion of these shapes and directions.

From page 7-8   7-6

No, this type of crater crack (✯) is not a transverse crack. Its general shape should give you a clue to its name - STAR crater crack.

However, another of the crater cracks *was* called a transverse crack. Turn ahead to page 7-10 and refresh your memory.

From page 7-3

You are *partially* correct in believing that a crater crack might be found at the beginning of a completed weld. The appearance of a crater crack here would be due to improper use of the heat source.

However, a crater crack would, or could, appear at the end of the weld or any stopping point in between.

Turn ahead to page 7-12 for further explanation.

From page 7-12   7-8

Yes, it is a definite fact that crater cracks *do* have different shapes and directions as indicated by the drawings on page 7-3. Here is a repeat of one of those drawings showing a crater crack.

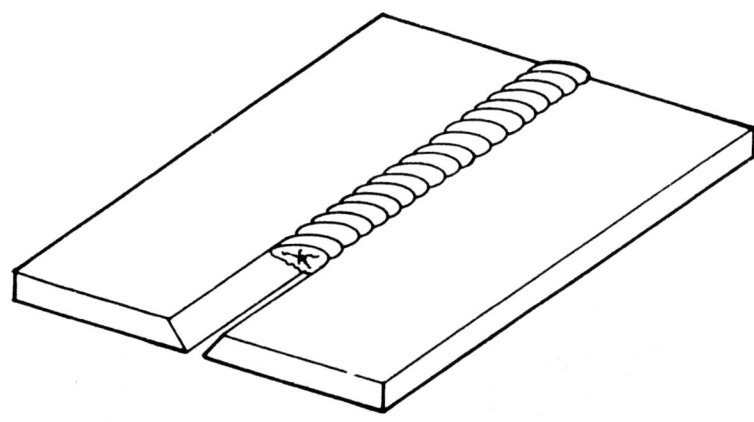

The above type of crater crack is given a name taken *from its general shape*.

What is this type of crater crack called?

Transverse .... **Page 7-6**
Star .... **Page 7-10**
Longitudinal .... **Page 7-13**

From page 7-3

You are right in thinking that a crater crack might occur somewhere between the start of the weld and the end of the weld. A temporary stop somewhere along the weld and an improper welding technique could cause such a crater crack.

HOWEVER, this answer is only *partially* correct, because the same reasoning applies to the beginning of the weld and the end of the weld.

For the completely correct answer, turn ahead to page 7-12.

From page 7-8 7-10

Correct! The rough STAR shape (✵) provides the clue to the name of this particular type of crater crack. Three types of cracks have been pictured. Shown below is a TRANSVERSE crater crack.

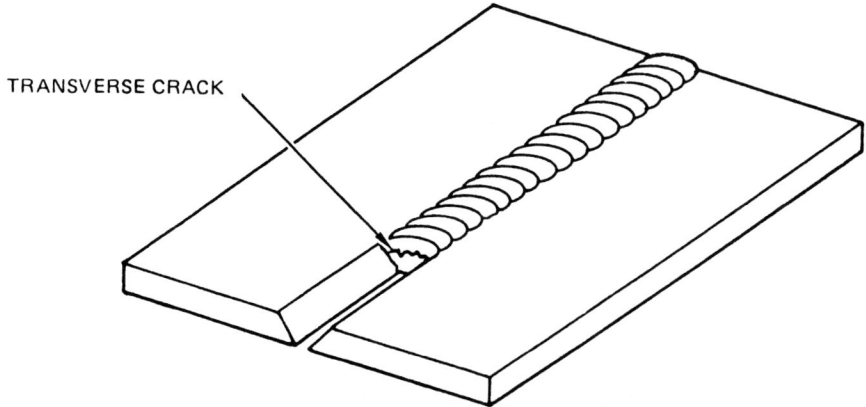

The reason for the name given this type of crack might not be as obvious as the reason for the name "star." But there *is* a definite reason for calling this kind of a crack a transverse crack. The reason can be found in the first part of the word - TRANSverse.

Trans simply means *across*. Recall the word *trans*ocean. It means *across* the ocean. *Trans*atlantic means *across* the Atlantic.

Take another look at the transverse crater crack above. A transverse crater crack is a crack that:

penetrates deep into the weld .... **Page 7-14**
runs across the crater .... **Page 7-17**

From page 7-17

Excellent! You have correctly determined that the first part of the word provides the key to the characteristics of a *long*itudinal crater crack. A longitudinal crater crack goes in the long direction of the weld, not across the weld or in a star shape.

So what's in a name? In these cases, a lot's in a name.

LONGitudinal - Goes in the *long* direction of the weld.

Let's review crater cracks:

- Star - A rough star-shaped crack.

- TRANSverse - Trans or across the weld.

- LONGitudinal - Goes in the long direction of the weld.

The terms *transverse* and *longitudinal* can be applied to any weld cracks that occur in the directions (across or long) indicated by those names.

Although a transverse crater crack is limited to the area of the crater and is caused by improper heat-source control, any crack that runs across the weld, regardless of the cause, is called a *transverse* crack.

The same holds true for a longitudinal crack. Any crack that parallels the direction of the weld bead is a *longitudinal* crack.

Turn ahead to page 7-16 for a discussion of stress cracks.

From page 7-3   7-12

Right. In a completed weld, a crater crack might be found at any one, or possibly all, of the following locations:

- At the end of the weld

- At the beginning of the weld

- Somewhere *between the beginning and the end* of the weld

If the heat source is improperly used at the *beginning* of the weld, a crater crack can occur. If the heat source is improperly handled at the *end* of the weld, a crater crack can occur. This stopping point might be at the very end of the metal to be welded, or it might occur before the weld is fully completed (a temporary stop). If the weld is not properly fused on restarting, the crater crack will remain.

What about the shape of these crater cracks?

They are all shaped alike .... **Page 7-5**
They can have different shapes and directions .... **Page 7-8**

From page 7-8   7-13

You chose to call the crater crack shaped like this (✲) a longitudinal crater crack. That name is not correct. A careful look at the shape of the crack should tell you that it is a STAR crater crack.

A type of crater crack *is* termed longitudinal, and it will be discussed shortly. But for now, turn back to page 7-10.

From page 7-10 7-14

A transverse crater crack might or might not penetrate deep into the weld, but that answer isn't correct. Remember that we have talked about the first part of the word *trans*verse. The first part means across. A transverse crater crack, then, is a crack which runs *across* the crater.

Turn ahead to page 7-17.

From page 7-17

The word longitudinal does not tell you that this type of crack occurs in a straight line; however, it does tell you that the crack occurs in the long direction of the weld bead.

Turn back to page 7-11 and see why.

From page 7-11   7-16

Stress Cracks

Stress cracks in welds are the result of stresses created during the cooling of a restrained (rigid) structure.

When the two pieces below were welded, the ends were joined and heated until they were molten. The molten material welded together and was left to solidify, joining the two pieces as one. As the molten metal cooled, it began to shrink, which created stress. The pieces were not restrained, so the stress relieved itself by pulling the two pieces up to form a "bow" rather than to crack. The "bow" can be undesirable and therefore may pose a problem. However, if each end of the welded part had been restrained by clamps, the metal would not have been able to relieve shrinkage stress by bowing and might have cracked.

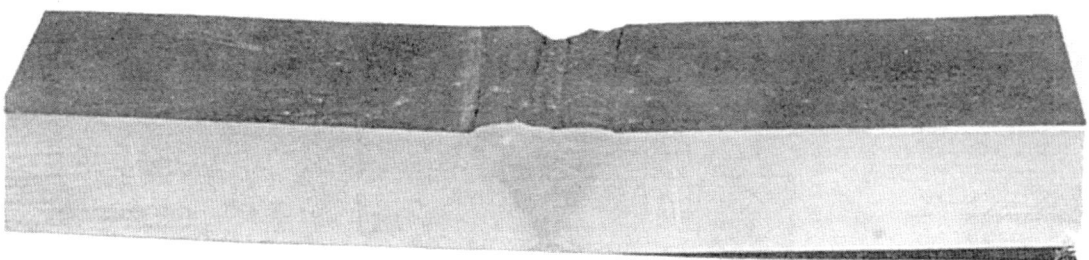

Turn ahead to page 7-18.

From page 7-10

Right! *Trans*verse means to go across, and a transverse crater crack goes *across* the crater from side to side.

Star and transverse cracks have been discussed. Now, for the third type of crater crack that might be caused by improper use of the heat source - LONGitudinal crater cracks.

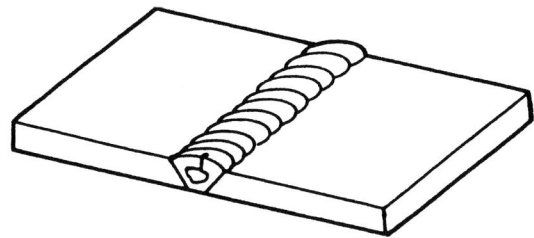

Once again, the name of the crack plays an important part in picturing and remembering the type of crack. The name "star" refers to the rough star-shaped crater crack. The term *trans*verse tells you that this type of crack goes across the weld.

On the other hand, the word longitudinal tells you that longitudinal crater cracks:

occur in the long direction of the weld . . . . Page 7-11
occur in a straight line . . . . Page 7-15

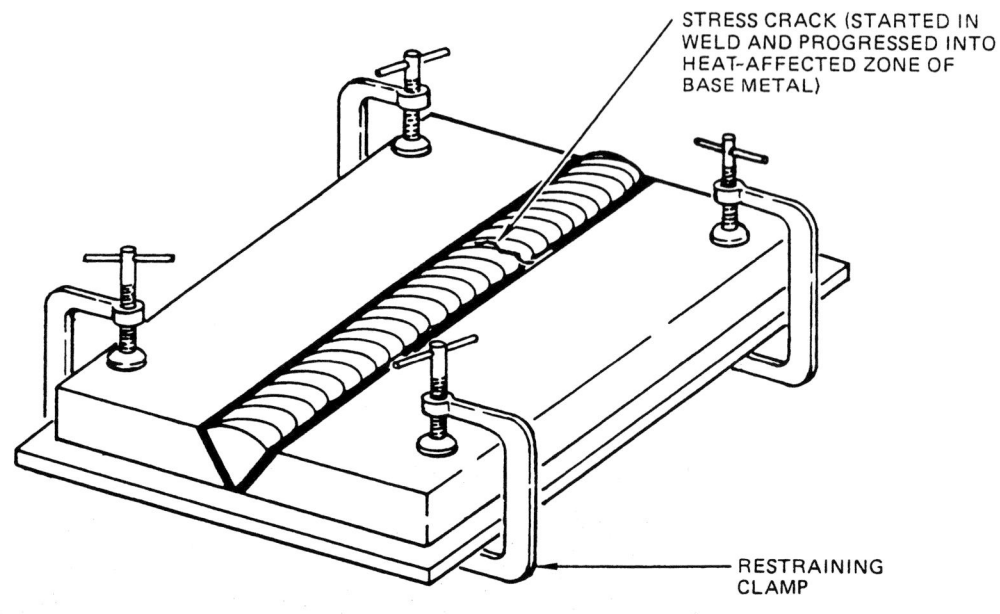

Stress cracks might occur anywhere along the weld bead, starting in the weld bead and working their way into the heat-affected zone of the base metal. These cracks usually occur transversely in a single-pass weld and longitudinally in a multiple-pass weld.

In a weld allowed to bow, the risk of stress cracks is avoided. However, welding two stationary pieces is necessary to minimize distortion in certain circumstances. An instance in which a stationary weld is unavoidable might be the welding of a catwalk which is already firmly attached to another structure.

A crack in a single-pass weld, such as the one just described, would usually occur:

transverse to the weld . . . . **Page 7-20**
in a longitudinal direction . . . . **Page 7-22**

From page 7-23   7-19

Porosity

Porosity, you will remember from the section on ingots, is entrapped gas. The same action that occurred in the ingot occurs on a smaller scale in the molten weldment - entrapped gas tends to rise toward the surface. If any of this gas remains entrapped in the weld, it is (as in the case of the ingot) called POROSITY. In fact, a weld *is* a cast structure.

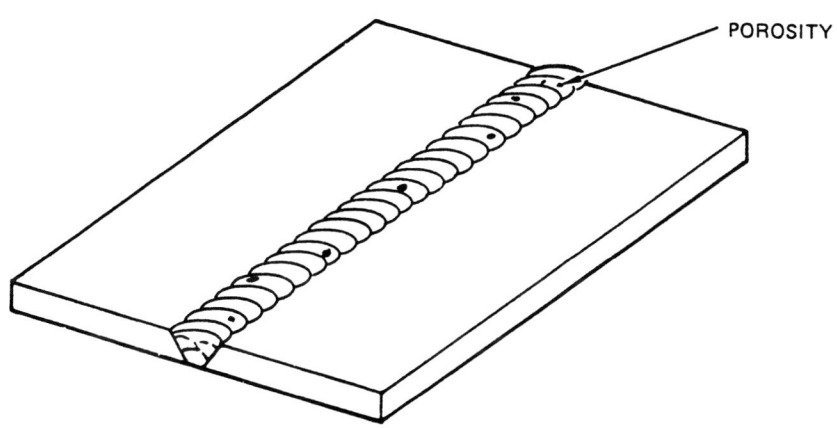

Entrapped gas would look like which of the following?

Ragged, irregular discontinuities . . . . Page 7-25
Round, or nearly round, discontinuities . . . . Page 7-27

From page 7-18

Yes. The stress crack would *usually* occur transverse (across) to the weld. Notice we said "usually." The pattern is not exact. The stress crack could occur in a longitudinal or nearly longitudinal direction, even in a single-pass weld.

A crack will often progress from the weld into the "heat-affected zone" (HAZ) of the base metal. The heat-affected zone is a narrow part of the base metal on each side of the weld. That part of the base metal is altered to some degree by the heat of the welding process.

Turn to the next page.

From page 7-20   7-21

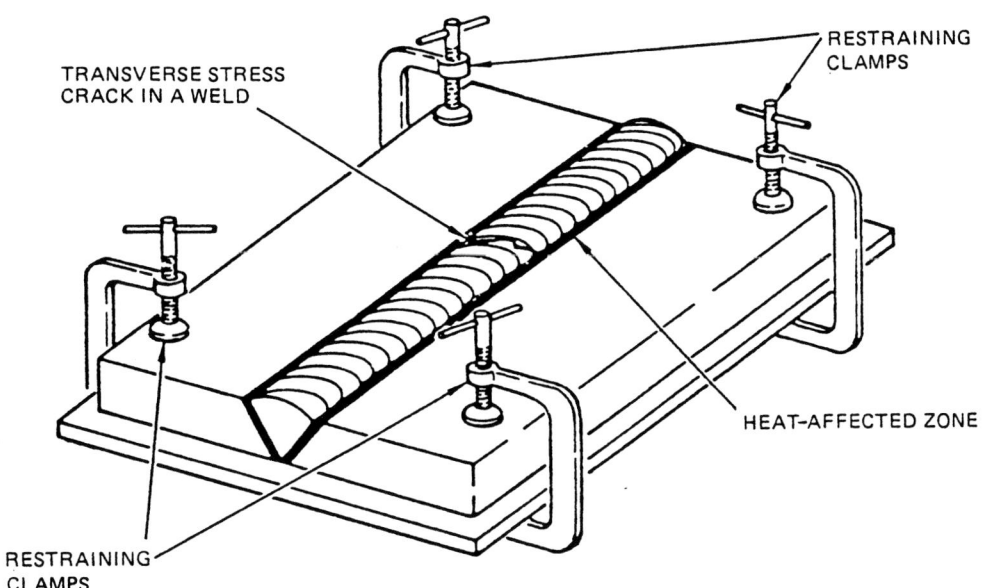

You can see by the above picture that the transverse crack has extended into the heat-affected zone of the base metal.

However, some cracks run in the same direction as the weld. What are these cracks called?

Longitudinal .... **Page 7-23**
Diagonal .... **Page 7-24**

From page 7-18   7-22

If two stationary (restrained) pieces were welded by a single pass and the stresses cause a crack, the crack would *usually* occur in a transverse, not longitudinal, direction.

Turn back to page 7-20.

Of course, a crack that runs in the same (long) direction as the weld is a *long*itudinal crack. It would look something like the ones pictured below.

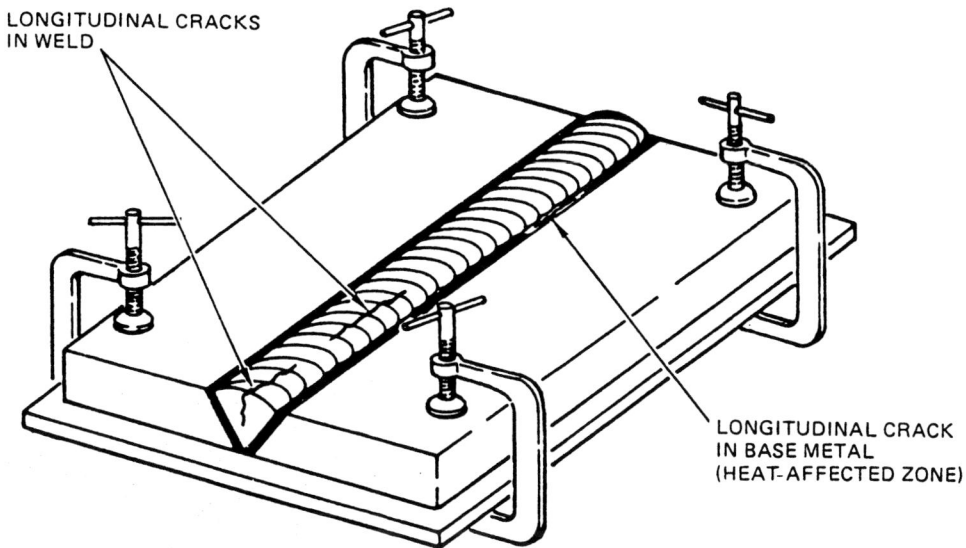

In summary, stress cracks in or near welds can and do occur in the following patterns:

- *Trans*verse cracks go *across* the weld.
- *Long*itudinal cracks occur in the direction *of* the weld.

Turn back to page 7-19 where another discontinuity that occurs in welds is discussed.

From page 7-21

Cracks that occur in the same direction as the weld bead are called *longitudinal* cracks. The word "diagonal" isn't very descriptive of a crack that runs in the *long* direction.

Turn back to page 7-23.

From page 7-19

No, porosity would not appear as ragged, irregular discontinuities. Porosity is entrapped gas. As was stated in the discussion on ingots, porosity has a bubble shape - round or nearly round.

Turn ahead to page 7-27.

From page 7-28   7-26

Of course. These impurities rise toward the top of the molten weldment. When these impurities reach the top of the puddle, they harden and form a crust of *slag*. This slag must be completely cleaned away before the welder makes another pass.

Much of this *slag* that is not cleaned off will probably be trapped (included) in the next layer of metal. As a result, SLAG INCLUSIONS will occur in the weld bead.

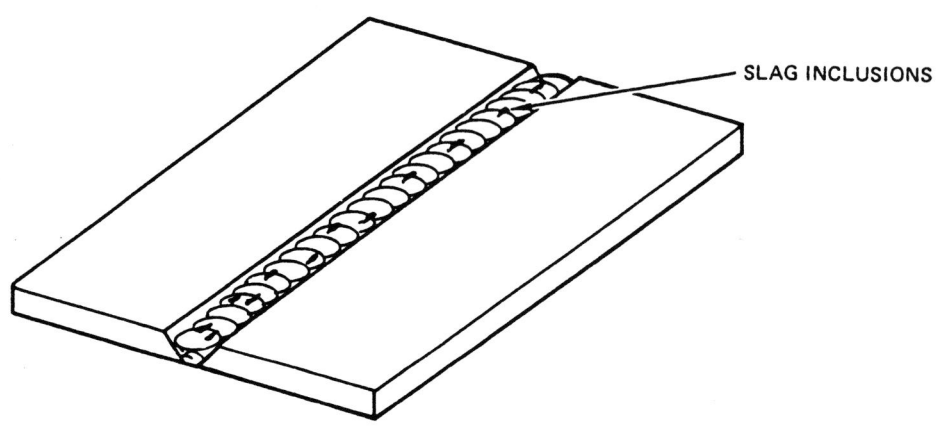

You can see from the above picture that slag inclusions in a weld:

occur only in one direction .... **Page 7-29**
have no definite direction .... **Page 7-32**

From page 7-19

Right. Porosity in a weld would appear as round, or nearly round, discontinuities. Porosity may be open to the surface, or it may be subsurface, depending on whether the gas was trapped by the solidifying metal.

Another discontinuity that might be found in welds is similar to a nonmetallic inclusion, but it is called a slag inclusion.

Turn to the next page.

Slag Inclusions

You will recall that slag in the original steel-making process was a source of unwanted impurities within the steel. Slag inclusions in welds are the same thing - unwanted impurities within the weldment.

Slag inclusions can occur during SMAW arc welding. As the electrode melts, its oxide coating partially melts and mixes with the molten metal, providing needed protection from impurities in the surrounding air.

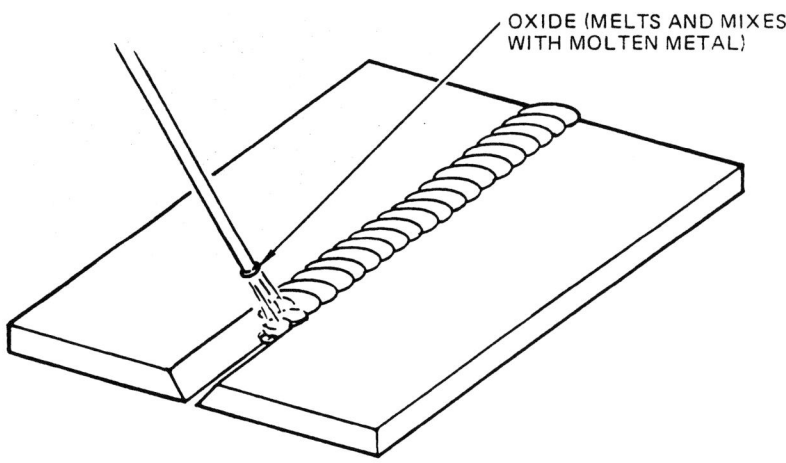

These oxide impurities react as nonmetallic impurities did in a molten ingot.

They rise toward the top of the molten metal .... **Page 7-26**
They all remain in the weld .... **Page 7-30**

Your choice that *slag inclusions* occur in only one direction is not correct. Slag inclusions occur in many positions.

Turn ahead to page 7-32.

From page 7-28

You have answered that all slag inclusions remain in the weldment. That isn't so. It was stated that slag inclusions react as nonmetallic inclusions did in an ingot. In other words, slag inclusions rise toward the surface of the molten metal.

Turn back to page 7-26.

From page 7-33

Although some tungsten inclusions might occur at the surface of a weld, most often they are *subsurface* discontinuities and *are not* open to the surface.

Turn ahead to page 7-36 for a look at these tungsten inclusions.

From page 7-26

Correct. Slag inclusions in a weld do not have a definite or specific direction. They were in the act of rising to the surface, but the metal solidified before they reached the surface, trapping them in various positions.

END VIEW

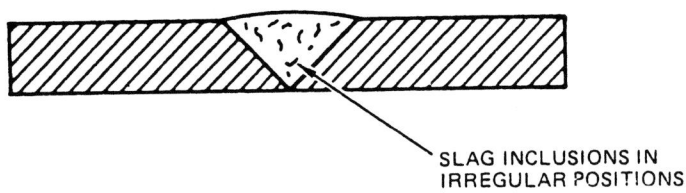

SLAG INCLUSIONS IN IRREGULAR POSITIONS

Turn to the next page.

From page 7-32

Tungsten Inclusion

There is another type of inclusion that might be trapped in a weld. Excessive current during tungsten arc welding (GTAW) can cause the tungsten electrode to melt. When this tungsten is deposited in the weld, a "tungsten inclusion" is created.

The name itself tells you that *tungsten inclusions* are:

open to the surface .... **Page 7-31**
subsurface .... **Page 7-36**

From page 7-36

The discontinuities that result when melted bits of the electrode are deposited in a weld are called *tungsten inclusions*. So, the electrodes are made of tungsten - not oxide. Oxide can cause a discontinuity, but it would be called a slag inclusion. Remember?

Turn ahead to page 7-39.

From page 7-37

You will *not* find lack of penetration present in the top of the weld. If this condition is present, the molten metal has not fused (welded) with the base metal. LACK OF PENETRATION, if it occurs, will be found in the root of the weld.

Turn ahead to page 7-38.

From page 7-33   7-36

Good. The word "inclusions" in tungsten inclusions tells you that these discontinuities are subsurface. The inclusions are pictured below.

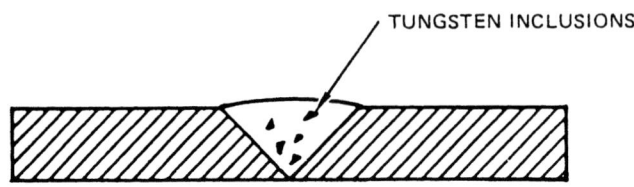

From the picture above, you can see that the inclusions are subsurface and, like slag inclusions, have no definite direction.

The discontinuities pictured above are the result of bits of the electrode being deposited in the molten metal during the welding process.

The electrode was made of:

oxide .... **Page 7-34**
tungsten .... **Page 7-39**

From page 7-39   7-37

Lack of Penetration

Inclusions, cracks, and porosity are not the only discontinuities to look for when inspecting welds. Bad welding technique can result in a discontinuity called "lack of penetration."

Lack of penetration, or incomplete penetration, is exactly what the name says - a failure of the molten metal, created by the welding process, to fuse with the parent or base metals to be joined during the first weld pass (known as the "root" pass).

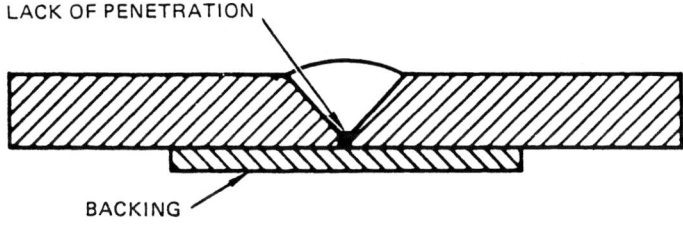

The situation shown above might have been caused from too much speed during the welding process, or it might have been caused by one of several other reasons.

Whatever the cause, when lack of penetration is present:

it occurs at the top of the weld .... **Page 7-35**
it occurs at the root of the weld .... **Page 7-38**

From page 7-37 7-38

Lack of penetration is a discontinuity which can occur at the root of a weld, of course. For one reason or another, the molten metal did not penetrate to the back, or bottom, of the weld joint.

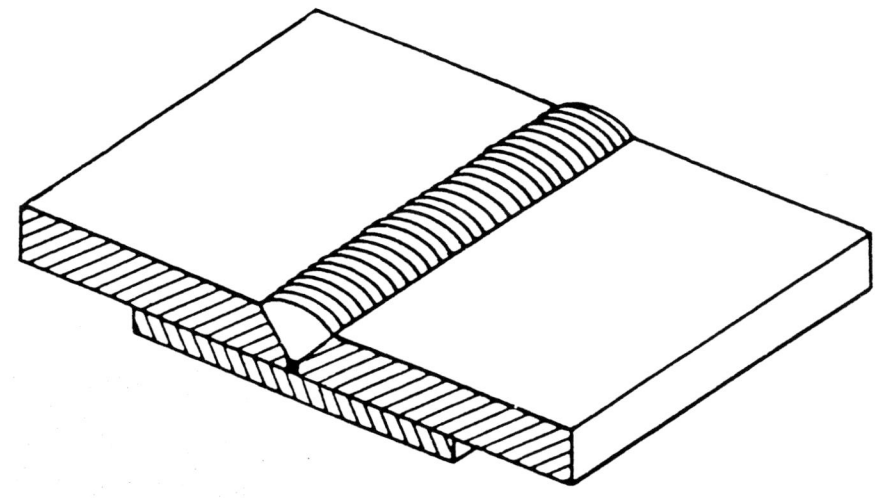

The above discontinuity results from:

lack of backing .... Page 7-40
root condition .... Page 7-43
lack of penetration .... Page 7-46

From page 7-36

Certainly. Tungsten inclusions get their name from the contaminating metal that caused them. The electrode is made of *tungsten*.

Turn back to page 7-37.

From page 7-38 7-40

The term for the discontinuity pictured on page 7-38 is not lack of backing. Part of the problem is that the molten puddle did not penetrate to the backing, but the word backing has nothing to do with the name of the discontinuity.

Turn back to page 7-38 and select another answer.

Bad guess. Nothing has been said (and nothing will be said) about line of demarcation. It has nothing to do with our welding discussion. Parent metal, however, is and has been a vital part of the discussion. The parent metal is the metal which is being welded together. You now know what the correct answer is, but turn back to page 7-46 for another look at LACK OF FUSION before proceeding.

Right! Lack of fusion is a failure of the weld to fuse with the parent metal. Sometimes the lack of fusion might occur between the weld passes. In either case, the discontinuity is termed LACK OF FUSION.

It occurs:

at the root of the weld .... **Page 7-44**
farther up in the weld than the root .... **Page 7-50**

From page 7-38   7-43

The discontinuity on page 7-38 does occur at the root or bottom of a weld, but it is not called "root condition."

Turn back to page 7-38 and select another answer.

From page 7-42

Lack of fusion occurs at the root of the weld? NO. A discontinuity that occurs at the root of the weld does have the word "lack" in its name, but it is not lack of *fusion*. Lack of fusion is a failure of the weld to fuse with the parent metal, and it might be caused by a failure of the weld passes themselves to fuse. This particular discontinuity occurs farther up in the weld than the root, correct?

Turn ahead to page 7-50.

From page 7-50   7-45

Undercut

The last discontinuity for our welding discussion is called UNDERCUT. In an UNDERCUT, the welder melts away and flushes out some of the parent or base metal along the line of fusion on the top of the finished weld. Here is what it looks like.

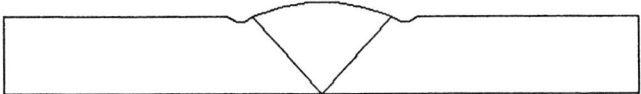

This discontinuity would be most readily seen by visual inspection; therefore, it is:

subsurface .... **Page 7-47**
open to the surface .... **Page 7-48**

From page 7-38   7-46

Correct! Lack of penetration is the right term for the discontinuity just pictured. It results from incomplete penetration into the parent metal or backing by the molten puddle.

Lack of Fusion

A similar condition, but one that occurs farther up than the root, is another "lack" known as "lack of fusion."

When a *lack of fusion* is found, it will look something like this:

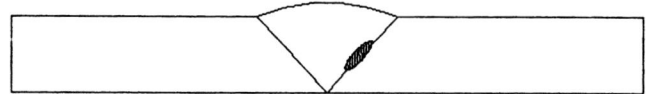

You can see that lack of fusion is a failure of the weld to fuse with the:

line of demarcation . . . . **Page 7-41**
parent metal . . . . **Page 7-42**

From page 7-45

No, an UNDERCUT is most readily seen by visual inspection. It would not be subsurface (beneath the surface), or it could not be inspected visually. It is, therefore, open to the surface.

Turn back to page 7-45, take another look at the picture of an undercut, and then turn to the page that the correct answer indicates.

Good, you realize that if *undercut* can be inspected visually, then it has to be open to the surface. It isn't hard to see. All that is necessary to inspect for undercut is to check the line of fusion; in other words, the meeting line of the weld and the parent metal. If there is undercut, some of the parent metal has been melted away.

Now, let's review the weld discontinuities in the order they were discussed.

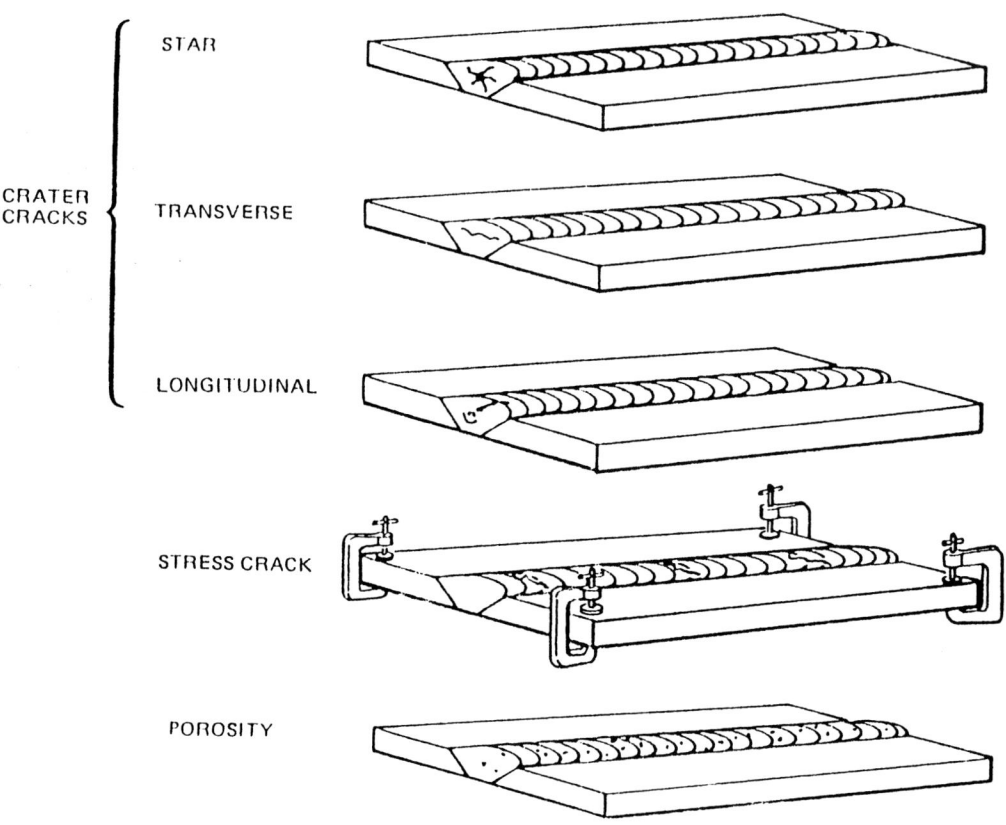

Turn to the next page for more weld discontinuities.

From page 7-48   7-49

Weld discontinuities continued.

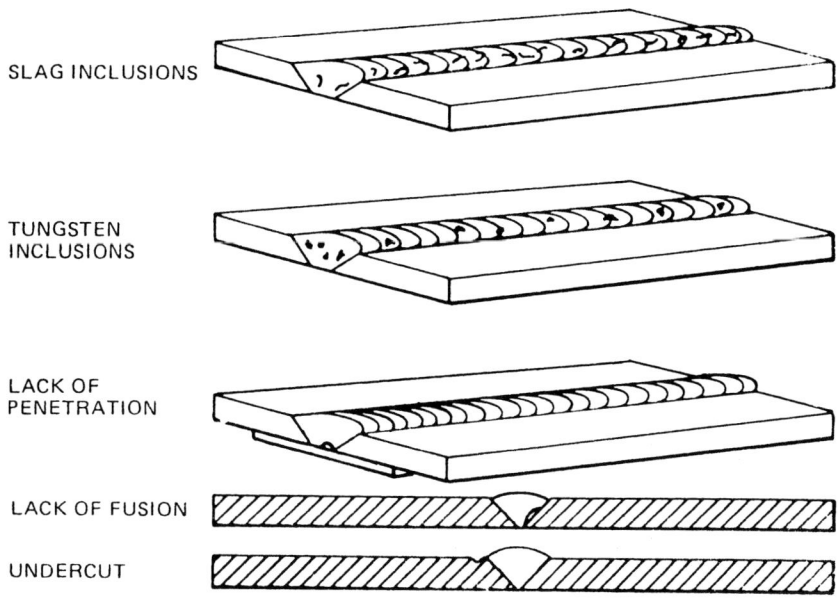

The publisher of this text, *PH Diversified, Inc.*, also markets a series of flawed specimen kits that contain a wide variety of welding discontinuities. The next four pages contain a detailed listing of welding discontinuities that are applicable to specific methods of NDT.

Turn ahead to page 7-51.

From page 7-42

Correct! Lack of fusion occurs farther up in the weld than the root. Perhaps the weld and base metal did not fuse, or maybe the weld passes themselves failed to fuse.

Turn back to page 7-45.

From page 7-49   7-51

FLAWTECH

ULTRASONIC EXAMINATION
FLAWED SPECIMEN KIT

Flaw Manufacturing Technology
A Division of **PH D**iversified, Inc.
5040 Highway 49 South
Harrisburg, NC 28075
Phone: (704) 455-1322
FAX: (704) 455-1323

10 SPECIMENS WITH 20 "REAL" FLAWS FOR LEVEL I & II TRAINING

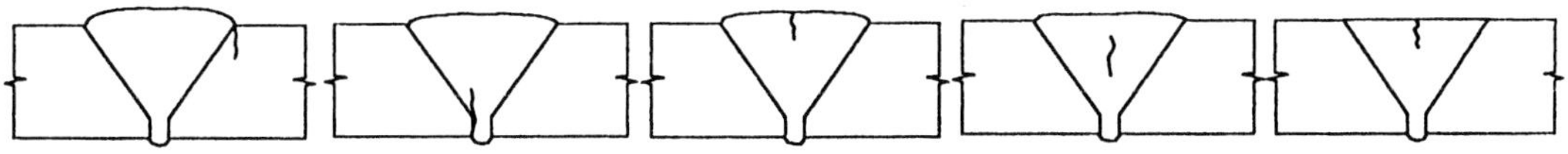

10. **Toe Crack** in Single Vee. MT/PT, UT, RT
12. **Root Crack** in Single Vee. MT/PT, UT, RT
14. **Centerline Crack**, in single Vee (surface). MT/PT, UT, RT
15. **Centerline Crack**, in Single Vee (sub-surface). UT, RT
16. **Circumferential Crack** in Single Vee, flush crown. MT/PT, UT

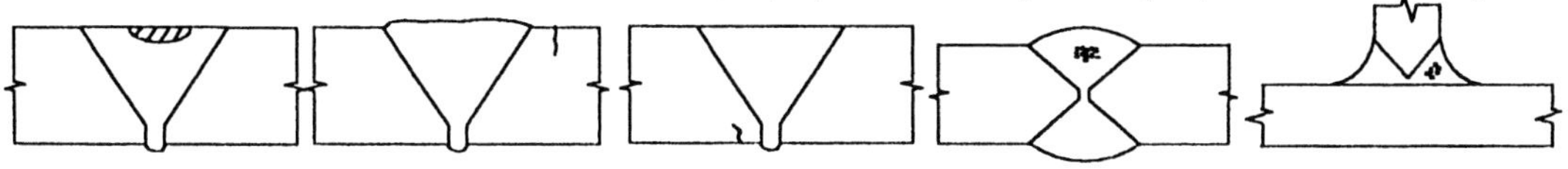

17. **Transverse Crack** in Single Vee, flush crown. MT/PT, UT
18. **Base Metal Crack** in Single Vee (top HAZ area). MT/PT, UT
19. **Base Metal Crack** in Single Vee (bottom HAZ area). MT/PT, UT
30. **Porosity** in Double Vee (sub-surface) UT, RT
31. **Porosity** In Fillet (sub-surface) UT, RT

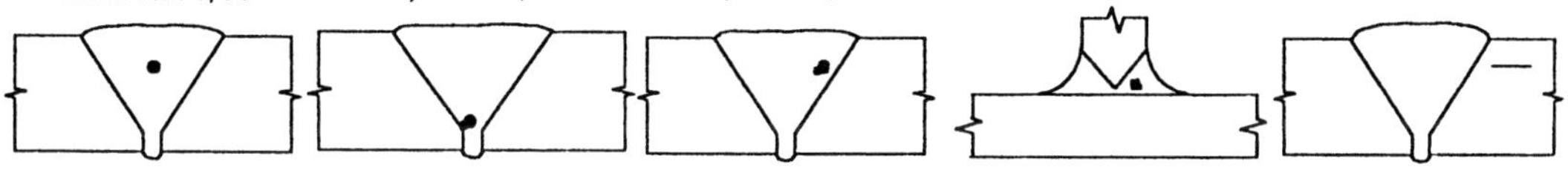

34. **Single Gas Pore** in Single Vee UT, RT
36. **Slag Inclusion** in Single Vee (bottom groove area). UT, RT
37. **Slag Inclusion** in Single Vee (top weld groove area). UT, RT
38. **Slag Inclusion** in Fillet (root area) UT, RT
50. **Lamination** in Single Vee (base metal) UT

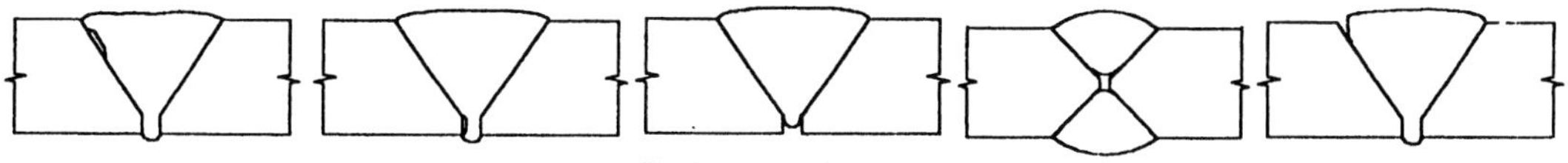

52. **Lack of Fusion** in Single Vee (crown area). UT
55. **Lack of Fusion** in Single Vee (surface at root). MT/PT, UT
56. **Incomplete Root Penetration** in Single Vee. VT, UT, RT
57. **Incomplete Root Penetration** in Double Vee UT, RT
59. **Incomplete Groove Weld** (crown area) VT, MT/PT, UT, RT

Drawings Not to Scale

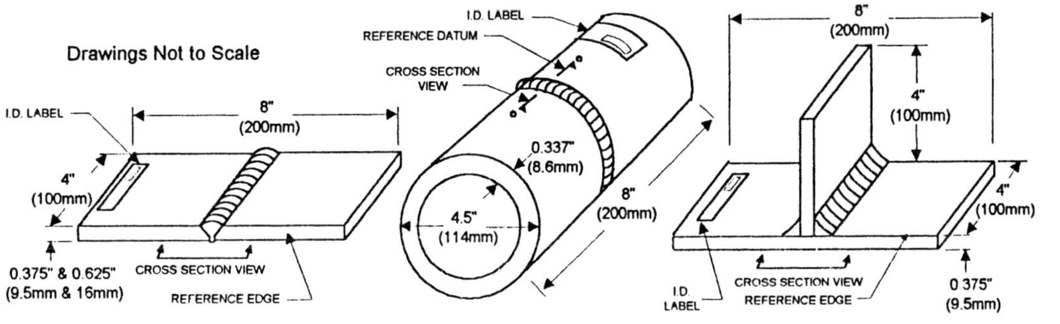

I.D. LABEL
8" (200mm)
4" (100mm)
0.375" & 0.625" (9.5mm & 16mm)
CROSS SECTION VIEW
REFERENCE EDGE

I.D. LABEL
REFERENCE DATUM
CROSS SECTION VIEW
0.337" (8.6mm)
4.5" (114mm)
8" (200mm)
I.D. LABEL

8" (200mm)
4" (100mm)
4" (100mm)
CROSS SECTION VIEW
REFERENCE EDGE
0.375" (9.5mm)

Turn to the next page.

F̲LAW̲T̲ECH

Flaw Manufacturing Technology

A Division of **PH D**iversified, Inc.
5040 Highway 49 South
Harrisburg, NC 28075
Phone: (704) 455-1322
FAX: (704) 455-1323

MAGNETIC PARTICLE/LIQUID PENETRANT FLAWED SPECIMEN KIT

10 SPECIMENS WITH 20 "REAL" FLAWS FOR LEVEL I & II TRAINING

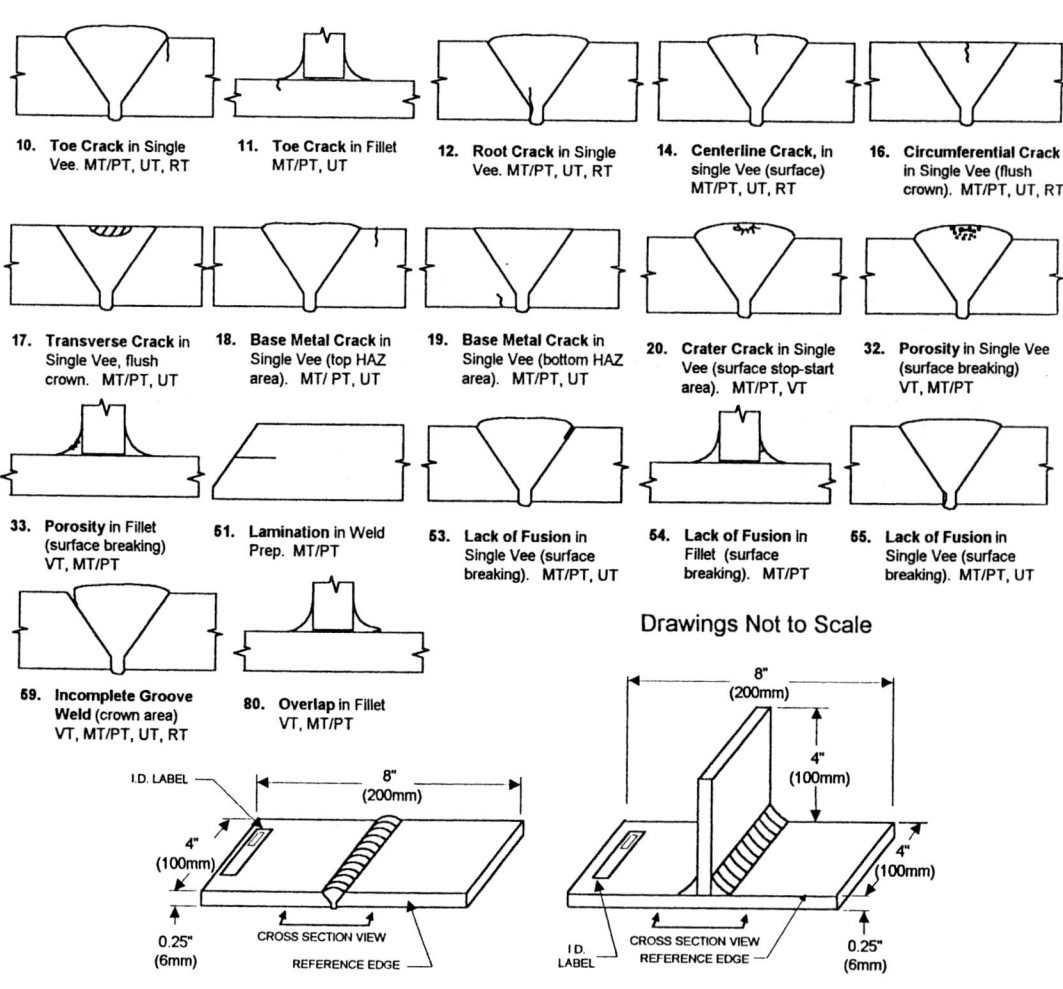

10. Toe Crack in Single Vee. MT/PT, UT, RT
11. Toe Crack in Fillet MT/PT, UT
12. Root Crack in Single Vee. MT/PT, UT, RT
14. Centerline Crack, in single Vee (surface) MT/PT, UT, RT
16. Circumferential Crack in Single Vee (flush crown). MT/PT, UT, RT
17. Transverse Crack in Single Vee, flush crown. MT/PT, UT
18. Base Metal Crack in Single Vee (top HAZ area). MT/PT, UT
19. Base Metal Crack in Single Vee (bottom HAZ area). MT/PT, UT
20. Crater Crack in Single Vee (surface stop-start area). MT/PT, VT
32. Porosity in Single Vee (surface breaking) VT, MT/PT
33. Porosity in Fillet (surface breaking) VT, MT/PT
51. Lamination in Weld Prep. MT/PT
53. Lack of Fusion in Single Vee (surface breaking). MT/PT, UT
54. Lack of Fusion in Fillet (surface breaking). MT/PT
55. Lack of Fusion in Single Vee (surface breaking). MT/PT, UT
69. Incomplete Groove Weld (crown area) VT, MT/PT, UT, RT
80. Overlap in Fillet VT, MT/PT

Drawings Not to Scale

Turn to the next page.

FLAWTECH

Flaw Manufacturing Technology

RADIOGRAPHIC EXAMINATION
FLAWED SPECIMEN KIT

A Division of **PH D**iversified, Inc.
5040 Highway 49 South
Harrisburg, NC 28075
Phone: (704) 455-1322
FAX: (704) 455-1323

10 SPECIMENS WITH 20 "REAL" FLAWS FOR LEVEL I & II TRAINING

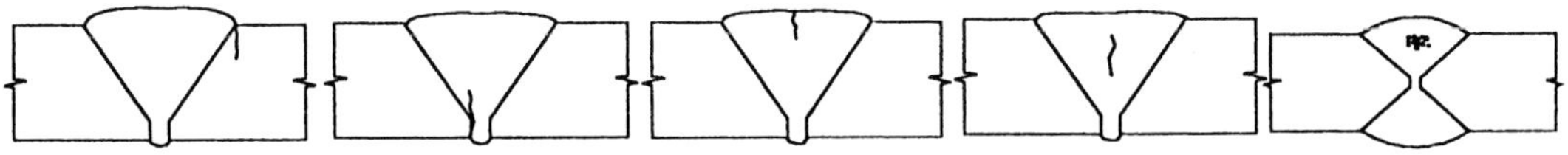

10. **Toe Crack** in Single Vee. MT/PT, UT, RT
12. **Root Crack** in Single Vee. MT/PT, UT, RT
14. **Centerline Crack**, in single Vee (surface) MT/PT, UT, RT
15. **Centerline Crack**, in Single Vee (sub-surface). UT, RT
30. **Porosity** in Double Vee (sub-surface) UT, RT

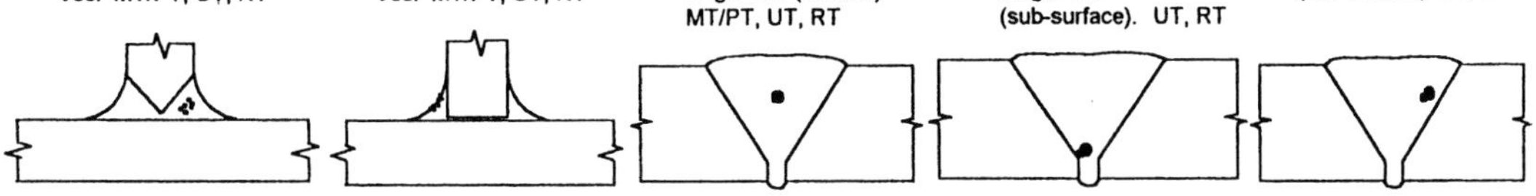

31. **Porosity** in Fillet (sub-surface). UT, RT
33. **Porosity** in Fillet (surface breaking) VT, MT/PT
34. **Single Gas Pore** in Single Vee UT, RT
36. **Slag Inclusion** in Single Vee (bottom groove area). UT, RT
37. **Slag Inclusion** in Single Vee (top weld groove area). UT, RT

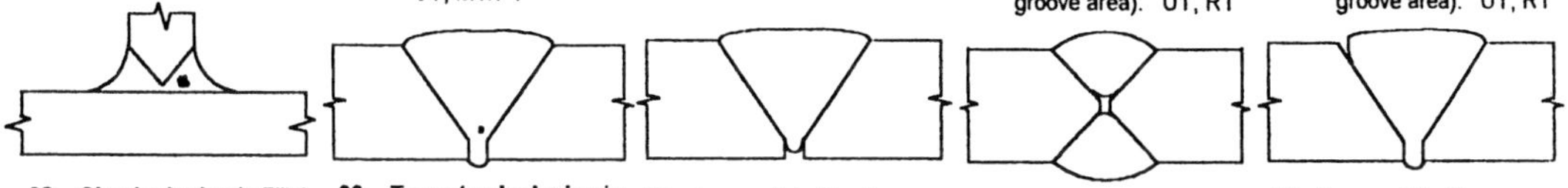

38. **Slag Inclusion** in Fillet (root area) UT, RT
39. **Tungsten Inclusion** in Single Vee (root area). RT
56. **Incomplete Root Penetration** Single Vee. VT, UT, RT
57. **Incomplete Root Penetration** in Double Vee UT, RT
59. **Incomplete Groove Weld** (crown area) VT, MT/PT, UT, RT

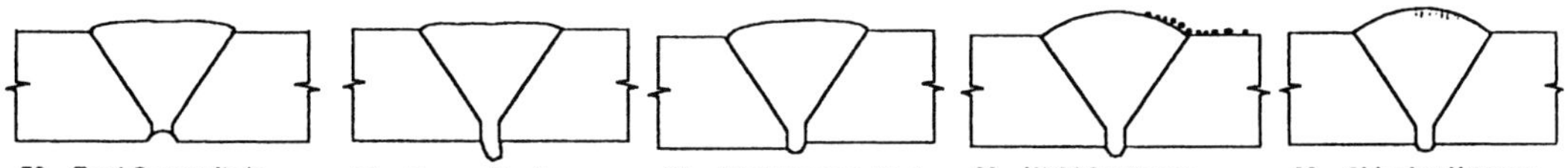

70. **Root Concavity** in Single Vee. VT, RT
71. **Excess Root Penetration** in Single Vee VT, RT
72. **Misalignment**, Root & Crown in Single Vee VT, RT
90. **Weld Spatter** on Single Vee VT, RT
92. **Chipping Hammer Marks** on Single Vee. VT, RT

Drawings Not to Scale

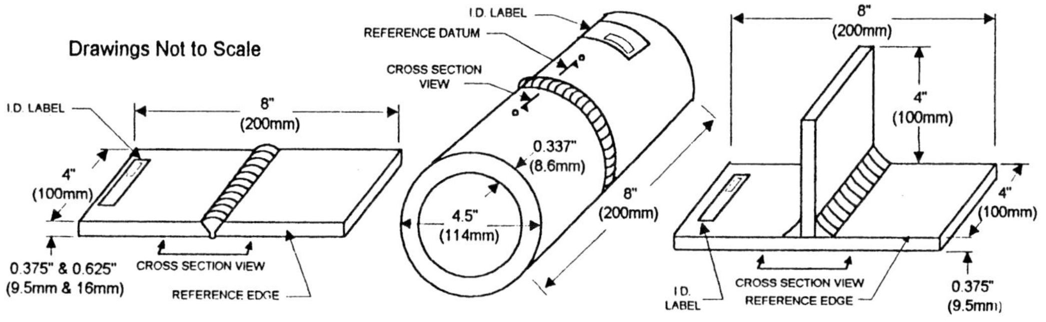

Turn to the next page.

FLAWTECH

Flaw Manufacturing Technology
A Division of **PH D**iversified, Inc.
5040 Highway 49 South
Harrisburg, NC 28075
Phone: (704) 455-1322
FAX: (704) 455-1323

VISUAL EXAMINATION
FLAWED SPECIMEN KIT

10 SPECIMENS WITH 20 "REAL" FLAWS FOR LEVEL I & II TRAINING

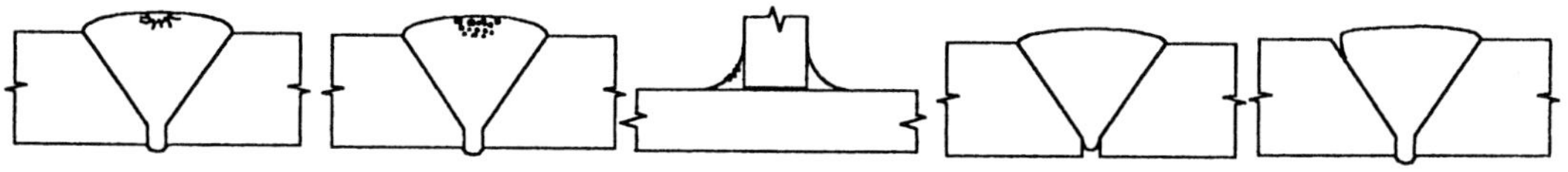

20. Crater Crack in Single Vee (surface stop-start area). MT/PT, VT
32. Porosity in Single Vee (surface breaking) VT, MT/PT
33. Porosity in Fillet (surface breaking) VT, MT/PT
66. Incomplete Root Penetration in Single Vee, VT, UT, RT
59. Incomplete Groove Weld (crown area) VT, MT/PT, UT, RT

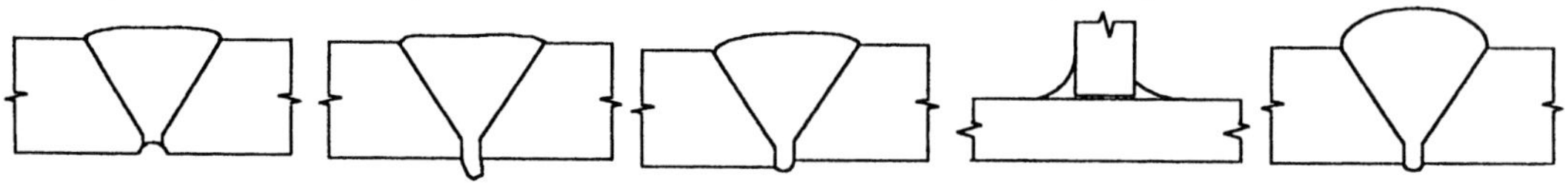

70. Root Concavity in Single Vee VT, RT
71. Excess Root Penetration in Single Vee VT, RT
72. Misalignment, Root & Crown in Single Vee VT, RT
73. Uneven Leg Length in Fillet VT
74. Excess Crown in Single Vee. VT

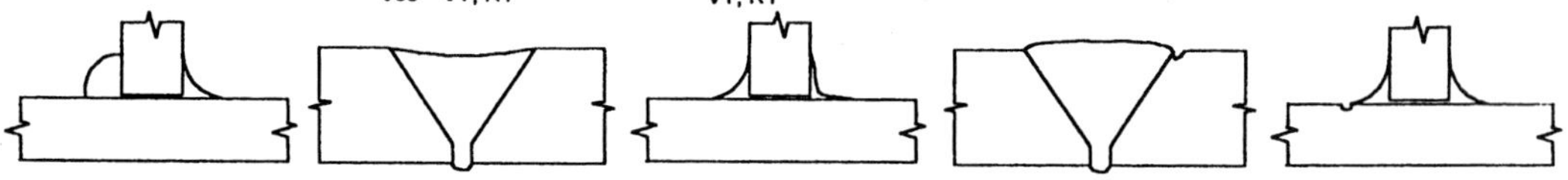

75. Excess Crown in Fillet VT
76. Concave Crown in Single Vee VT
77. Concave Crown in Fillet VT
78. Undercut in Single Vee VT
79. Undercut in Fillet VT

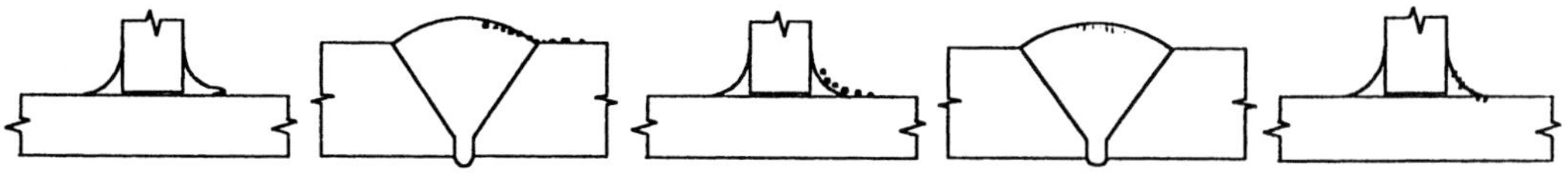

80. Overlap in Fillet VT, MT/PT
90. Weld Spatter on Single Vee. VT, RT
91. Weld Spatter on Fillet. VT, RT
92. Chipping Hammer Marks on Single Vee. VT, RT
93. Chipping Hammer Marks on Fillet. VT, RT

Drawings Not to Scale

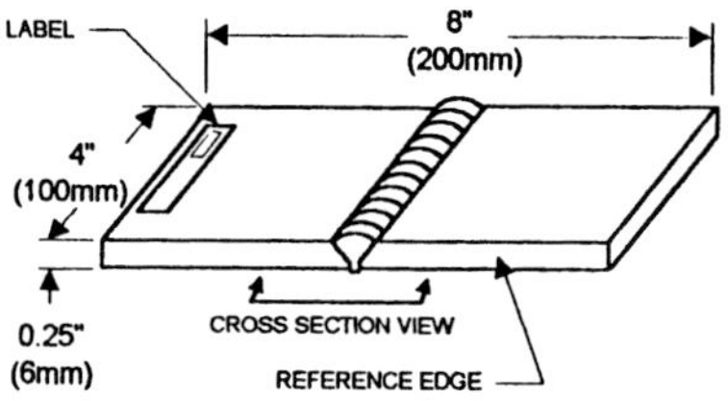

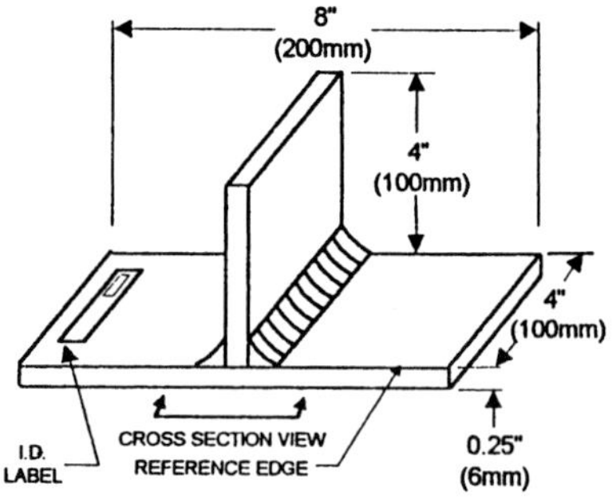

Turn to the next page for a chapter review.

CHAPTER REVIEW

_____ 1. As with other steel-working processes, welding can cause discontinuities within metal. The first ones we discussed were crater cracks. These are caused by the improper use of the heat source and assume several shapes. One shape gets its name from its _____ shape.

    A.   ball-like
    B.   star-like
    C.   slag-like
    D.   crack-like

_____ 2. Crater cracks can occur:

    A.   at the beginning of the weld.
    B.   at the end of the weld.
    C.   anywhere in between.
    D.   at any of the above locations.

_____ 3. Crater cracks are classified as one of the following three types:

    A.   ball, star or transverse.
    B.   star, crack or longitudinal.
    C.   star, longitudinal or transverse.
    D.   star, demarcation or longitudinal.

____ 4. _____ crater cracks run _____ the weld.

   A.   Longitudinal, across
   B.   Star, clockwise
   C.   Transverse, parallel to
   D.   Transverse, across

____ 5. _____ crater cracks run _____ the weld.

   A.   Longitudinal, with
   B.   Star, clockwise
   C.   Transverse, with
   D.   Longitudinal, across

____ 6. Any type of crater crack *could* occur:

   A.   at the beginning of the weld.
   B.   at the end of the weld.
   C.   in between the beginning and the end.
   D.   at any of the above.

___ 7. Another crack that can occur in welds is due to the welded pieces being restrained. All welding builds up a certain amount of _____ within a metal. In a rigidly-restrained parent material, it might relieve itself as a _____.

    A.   stress, crack
    B.   strain, tear
    C.   stress, die
    D.   strain, star

___ 8. Cracks caused by stress usually occur _____ to the weld. But since the heat-affected zone is a temporarily weak area, _____ cracks could occur in the parent metal.

    A.   longitudinal, longitudinal
    B.   transverse, stress
    C.   longitudinal, stress
    D.   transverse, transverse

___ 9. Stress cracks which occur in the heat-affected zone of the parent metal would probably extend in the same direction as the weld or _____ to the weld.

    A.   longitudinal
    B.   transverse
    C.   clockwise
    D.   counterclockwise

From page 7-57   7-58

_____ 10. Cracks are not the only type of discontinuity we might find in a weld. Another discontinuity, _____, is caused by entrapped gas.

   A. non-metallic inclusion
   B. slag
   C. porosity
   D. lack of penetration

_____ 11. If the welder should fail to remove all the slag between each weld pass, _____ may occur.

   A. slag inclusions
   B. porosity
   C. lack of penetration
   D. undercut

_____ 12. Another improper welding technique - excessive welding current during tungsten arc welding - causes a _____ discontinuity termed _____.

   A. surface, tungsten exclusion
   B. subsurface, tungsten inclusion
   C. surface, tungsten slag
   D. subsurface, tungsten porosity

From page 7-58

13. If excessive welding current is used when welding with the GTAW process, a discontinuity often caused by the melting of the electrode will consist of what type of metal?

   A. Star
   B. Slag
   C. Tungsten
   D. Arc

14. Cracks, inclusions and porosity are among the several types of welding discontinuities. One type of discontinuity occurs when the weld fails to fuse with the parent metal at the root of the weld. It is called:

   A. lack of weld.
   B. lack of penetration.
   C. lack of fusion.
   D. undercut.

15. _____ occurs at the root or bottom of the weld. Another discontinuity also occurs because of a failure to fuse properly and can be found farther up in the weld then the root and is known as _____.

   A. Lack of fusion; lack of weld.
   B. Lack of tungsten; lack of fusion.
   C. Lack of fusion; lack of penetration.
   D. Lack of penetration; lack of fusion.

16. _____ occurs when the weld does not fuse with the _____ metal or when there is a failure of the weld passes themselves to fuse.

   A. Lack of penetration, main
   B. Lack of fusion, parent.
   C. Lack of penetration, parent.
   D. Lack of fusion, main.

17. Lack of penetration and lack of fusion are due to improper welding technique, as are most welding discontinuities. Poor welding techniques can also lead to cutting of the metals to be joined which is called:

   A. undercut.
   B. back cut.
   C. trough cut.
   D. cut back.

18. When a welder melts away the parent metal along the line of fusion, he is creating a discontinuity called _____. This discontinuity is visible to the visual examiner eyes and is, therefore, a _____ discontinuity.

   A. back cut, subsurface
   B. over cut, surface
   C. undercut, subsurface
   D. undercut, surface

From page 7-60   7-61

_____ 19. There are three types of crater cracks. One is called _____ because of its shape, another goes across or _____ to the weld, and the third runs in the direction of the weld and is called a _____ crack.

    A.   ball, star, transverse
    B.   star, crack, longitudinal
    C.   star, transverse, longitudinal
    D.   star, demarcation, longitudinal.

_____ 20. Two types of weld inclusions are:

    A.   slag and tungsten.
    B.   penetration and fusion.
    C.   longitudinal and transverse.
    D.   surface and subsurface.

Select the matching weld discontinuity from the sketches shown below.

_____ 21. Misalignment   _____ 25. Slag Inclusion
_____ 22. Undercut   _____ 26. Incomplete Root Penetration
_____ 23. Root Crack   _____ 27. Incomplete Groove Weld
_____ 24. Porosity   _____ 28. Root Concavity

Turn to the next page for answers to these review questions.

ANSWERS TO REVIEW QUESTIONS FOR CHAPTER 7

Question & Answer		Reference Page(s)
1.	B	7-6
2.	D	7-12
3.	C	7-3
4.	D	7-10
5.	A	7-11
6.	D	7-11
7.	A	7-16
8.	B	7-20
9.	A	7-23
10.	C	7-19
11.	A	7-26
12.	B	7-33
13.	C	7-33
14.	B	7-37
15.	D	7-38, 7-46
16.	B	7-46
17.	A	7-45
18.	D	7-45
19.	C	7-11
20.	A	7-28, 7-33

Turn to the next page.

Question & Answer	Reference Page(s)
21. D	7-51 thru 7-54
22. H	7-51 thru 7-54
23. A	7-51 thru 7-54
24. E	7-51 thru 7-54
25. B	7-51 thru 7-54
26. F	7-51 thru 7-54
27. C	7-51 thru 7-54
28. G	7-51 thru 7-54

Take a short break and then begin Chapter 8.

CHAPTER 8

OTHER ENGINEERING MATERIALS

We have concentrated on metals thus far. Several other material groups have undergone increased emphasis as valuable and important engineering materials. They include the following general categories:

- Plastics
- Composites
- Ceramics
- Powder Metallurgy

Although we didn't discuss the structure of a material, a brief general description will be given here. Understanding these concepts is necessary for us to appreciate the usefulness of these materials.

Turn to the next page.

From page 8-1

The ATOM IS THE SMALLEST AMOUNT OF AN ELEMENT THAT CAN EXIST. Carbon is an element, as is hydrogen, and each can exist as a single atom.

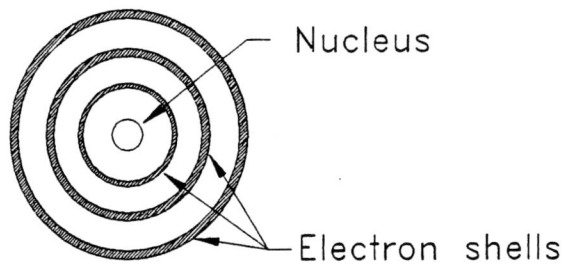

Water, written H_2O, is two hydrogen atoms attached to one oxygen atom. H_2O is a MOLECULE, WHICH IS THE GROUPING TOGETHER OF ATOMS BY CHEMICAL BONDS.

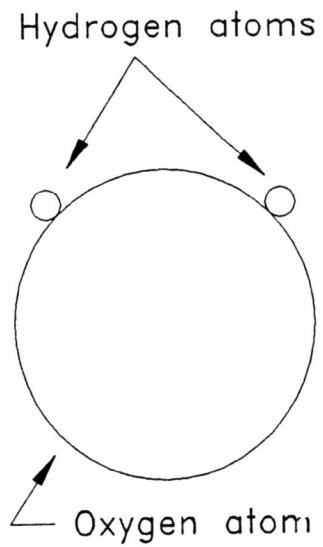

How would you define a molecule?

The smallest amount of an element .... Page 8-4
The grouping together of atoms .... Page 8-6

From page 8-6   8-3

In some cases, the molecular binding forces are strong. Have you ever stopped to think about the forces holding objects together, such as a pencil, the earth, or how about the forces holding the planets in orbit around the sun?

Turn ahead to page 8-6 and select one of the other answers.

From page 8-2

You have confused the definitions you just read. An ATOM is the smallest amount of an element that can exist and still be that element. However, the molecule is a grouping together of atoms. Refer to the figure below then turn ahead to page 8-6.

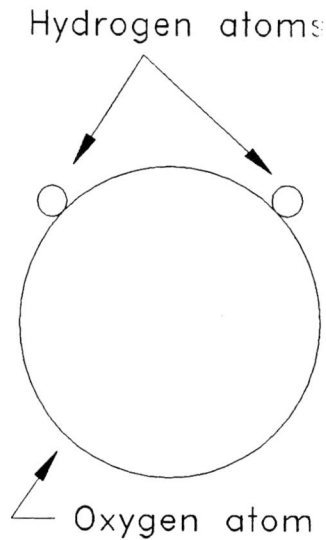

From page 8-6

You are partially correct. The air we breath contains gases, atoms, and molecules that are held together loosely. Yet some materials, like steel, are dense and hard, indicating that their atomic forces are strong.

Turn ahead to page 8-8.

From page 8-2   8-6

Great! You catch on quickly. A MOLECULE is the grouping together of atoms to form a substance. If further divided, that substance would result in individual atoms.

It is difficult to appreciate the small size of atoms. There are more than ten billion atoms in the smallest speck we can put under a microscope. Put another way, the atom is one million times smaller than the size of a human hair. More interesting is the fact that an atom is made of at least three smaller particles!

Nature assembles atoms into specific arrangements or molecules. We too can place atoms together in a way that is beneficial to a specific need. Engineers design the structure of a fighter aircraft wing based on the lightest and strongest materials they can manufacture. The BINDING FORCES of atoms and molecules keep materials together.

Would you say that the binding forces of all molecules are weak or strong?

Strong 8-3
Weak 8-5
Can't really say 8-8

As water freezes you can see the crystals form -- if you are patient enough to wait. On the other hand, if we pour water on ice we can actually hear the crystals breaking apart. Yet the ice is just water that is frozen into a solid.

As atoms form molecules, they can often grow crooked or out of "perfect" order. As a crystal grain grows, it will eventually meet other crystals growing next to it and form grain boundaries where they meet.

Sometimes impurities are caught in between the atoms as they group together. This creates fault lines in the atomic structure. It is these fault lines that jewelers use to cut diamonds.

Turn ahead to page 8-9.

Excellent! This is the best answer for the question. The air around us consists of weakly-bonded atoms and molecules. A rubber band can be stretched considerably, whereas a diamond can be cut only in a very specific way. Binding forces are very different for different materials.

We should begin to understand that the arrangement of specific atoms around other atoms creates substances like diamonds and water. *An ASSEMBLY of atoms to form structures is known as a CRYSTAL.*

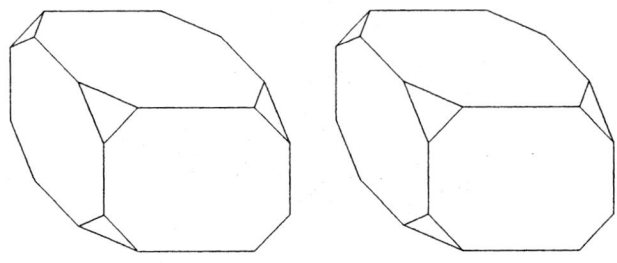

Turn back to page 8-7.

From page 8-7

The structure of a material determines many of its final characteristics. When we split a wooden log, we are careful to position our axe in just a particular orientation to the log so that the log splits easily. Grain direction, once again, is related to the strength of many engineering materials.

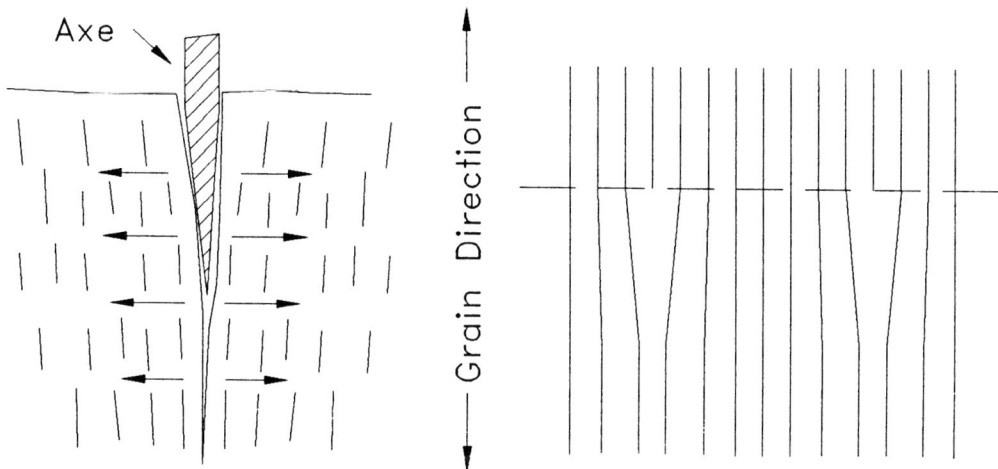

Grain direction can be related to the strength of a material.

True **8-10**
False **8-12**

From page 8-9

You bet! Grain direction and size plays a role in determining the mechanical properties of a material. Things such as strength, toughness, and hardness are related to the grain's direction and size.

Turn ahead to page 8-11.

The manner in which molecules and grains develop will have a large effect on the classification of a material. Metals have strong and well-organized structures. Engineering materials like plastics, composites, and ceramics may be either hard or soft in structure.

Scientists and engineers can actually produce materials that have the specific mechanical properties they need. For instance, that aircraft wing we mentioned earlier must be strong, flexible, able to withstand temperature changes, and lightweight. An internal combustion engine is cooled so that the metal it is made from does not overheat and fail. Ceramic materials are used in automotive engines today with a minimum concern for cooling, making them more efficient and lighter in weight. Let's look at each of these types of engineering materials.

Turn ahead to page 8-13.

Oh, but it can! Remember that diamond? Jewelers use those dislocations in the grain to break the stone into smaller pieces. That would not be possible if it were cut with the strength of the grain!

Turn back to page 8-10.

From page 8-11

Plastics

Plastic comes from the Greek word *plastikos*, which means "able to be molded." Plastics might best be defined as:

> MAN-MADE RESINS THAT FORM SUBSTANCES THAT CAN CHANGE SHAPE UNDER HEAT AND PRESSURE WHEN LINKED TOGETHER IN A VARIETY OF WAYS.

Individual molecules, or links, used to build the man made (or synthetic) resins are called *monomers*. When these monomers are linked together, through *polymerization*, a plastic is formed. This process is similar to the process of assembling the links of a chain (monomer) to form the chain (polymer).

Plastics can best be defined as:

molecules known as monomers .... **Page 8-14**
the linking of man made resins .... **Page 8-17**
another name for molds .... **Page 8-20**

From page 8-13

Not exactly. Plastics are formed by the linking of monomers to form chains or polymers of molecules. These are then called plastics.

Turn back to page 8-13 for a review and try the question again.

From page 8-17

We haven't talked about a thermomolecular plastic, so we need to read the answers more carefully. We have discussed thermoplastic and thermosetting plastics.

Turn ahead to page 8-19 and continue.

From page 8-19

Composites

Composite materials are commonly employed in today's engineered designs. A COMPOSITE MATERIAL is a MATERIAL MADE UP OF DISTINCT PARTS, and is usually ONE MATERIAL LAYERED ONTO ANOTHER, WITH EACH MATERIAL KEEPING ITS OWN IDENTITY. They offer diverse physical and mechanical properties and are generally very lightweight.

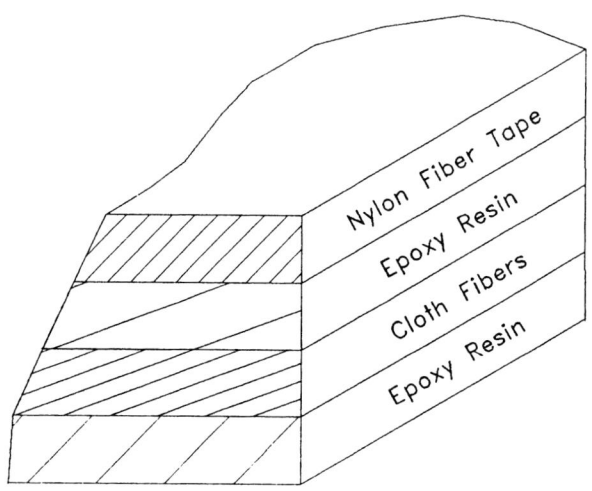

Turn ahead to page 8-18.

From page 8-13

Excellent! The linking of man-made resins through polymerization forms plastics. Plastics are generally of two types: *thermoplastic* or *thermosetting*.

Thermoplastics are formed when end-on-end chains are formed; therefore, they can be heated and remolded into a different shape quite easily.

On the other hand, thermosetting plastics are formed when monomers cross-link with one another. They are more rigid and cannot be remelted and remolded into shape after they have initially "set."

Plastics can generally be placed in two categories:

thermomolecular and polymerizing. . . . . **Page 8-15**
thermoplastic and thermosetting. . . . . **Page 8-19**

From page 8-16   8-18

For example, a metallic composite material used in an aircraft structure consists of a center tungsten wire coated with boron and treated with a silicate coating. By bonding these "fibers" to aluminum, they are made into a tape with a thickness of about 0.005 inch (0.13 mm). Aircraft structures are then made by bonding layers of the tape.

A composite is:

the same as a plastic. .... Page 8-21
made of materials that are layered, usually one atop another. .... Page 8-22

You are absolutely correct! The two main types of plastics we have discussed are thermoplastic and thermosetting. Thermoplastics include ABS used for helmets and appliances, and cellulose acetate butyrate used in the manufacture of telephone headsets and steering wheels.

Thermosetting plastics include urethane used for gears, o-rings, footwear, and upholstery; and phenolics used in buttons, toys, and poker chips. Plastics are also in soft and liquid form which serve as refuse bags and lubricants respectively. *Ceramoplastics* are also available and are defined as *heat-resistant plastics made by combining synthetic mica and glass*.

To further complicate matters, there are plastics that combine the thermoplastic and thermosetting molecular structure. All in all, engineers utilize the various polymerization processes available to manufacture plastics that meet specific applications.

Classification and description of discontinuities with plastics is difficult because of the various types, shapes, and forms. One can rest assured that opportunities abound for the nondestructive testing of such materials!

Turn back to page 8-16 as we discuss composites.

From page 8-13

Though we have mentioned that plastics are often molded into a shape for end use, this does not define plastic. Plastics are formed when monomers form chains through the process of polymerization.

Turn back to page 8-13 and try the question again.

Not by definition. Composites may well include plastic materials in their structure, but they are not generally considered plastics. The petrochemical industry utilizes fiberglass reinforced vessels (FRPs) as storage and processing tanks or vessels. Though referred to as FRPs, these are made by layering epoxy with glass (or cloth or paper) fibers in various directions to yield the proper structural shape and characteristics desired. One can see that such composites offer an endless array of uses and designs.

Turn to the next page.

From page 8-18   8-22

Correct! Layered materials constitute a composite structure. As we've said, this could be layers of epoxy followed by layers of glass fibers or many other combinations of materials.

As with plastics, NDT of composites is often limited to locating major *laminations* in the layers of materials. NDT can often locate a dense material within a less dense material, as in the tungsten wire in the tape we just discussed. Do you think there are materials, besides metal, that are neither plastics nor composites?

OK, what would you call them?

Plasti-metals .... **Page 8-25**
Ceramics .... **Page 8-26**

From page 8-26

Sorry, the question clearly asks, "Materials that are NEITHER plastics nor metals," and thermoplastic and thermosetting refer to the two main types of plastics.

Turn ahead to page 8-26 and make another choice..

From page 8-26

You are partially correct. SILICATES consist of the minerals clay, feldspar, silica, and bauxite. Ceramics are made from silicates. The more correct choice is ceramics.

Turn ahead to page 8-29.

From page 8-22

What the heck! You took a guess and, unfortunately, you were wrong. Now that you know the correct answer is ceramics, turn to the next page and continue.

From page 8-22　　　　　　　　　　　　　　　　　　　　　　　　　　8-26

Ceramics

CERAMICS are MATERIALS THAT ARE NEITHER PLASTICS NOR METALS. Ceramics are compounds of oxygen, nitrogen, and carbon that have been through high-temperature processing.

Minerals called SILICATES are found in abundance in the earth's crust. SILICATES consist of the minerals clay, feldspar, silica, and bauxite. Ceramics are made from silicates. By controlling the amounts and compositions of each silicate that is combined to make a specific ceramic, engineers can control the properties of the final ceramic product to satisfy a particular need.

Ceramics are useful in high-temperature applications, corrosive environments (such as electrical conductors) and, of course, dinnerware. Recall the ceramic engine that promises to be lighter and more fuel-efficient because of its heat-resistant properties.

Materials that are neither plastics nor metals are called:

thermoplastics .... **Page 8-23**
silicates .... **Page 8-24**
ceramics .... **Page 8-29**

From page 8-30

Casting is not correct. Remember, casting was discussed earlier as a process by which molten metal was poured into molds from ingots for final processing. No pressure or additional heat is applied to the molten metal.

P/M is a process in which powders of metals are made, then mixed with binding agents, pressed into the shape of a mold, then heated (or sintered).

Turn to the next page.

From page 8-30

Bravo! Seems you have a handle on what the powder metallurgy process is. Good. Let's look at the discontinuities with P/M.

Concerns for the powder metallurgy industry include, primarily, density variations throughout the part. This occurs mainly due to variations in punch (or press) pressures on the powder.

These density variations may result in a cracked part when it is ejected from the mold, though a service-related failure could also occur.

Turn ahead to page 8-31.

Correct! CERAMICS are MATERIALS THAT ARE NEITHER PLASTICS NOR METALS. Ceramics are compounds of oxygen, nitrogen, silicates, and carbon that have been through high-temperature processing.

Turn to the next page.

From page 8-29

Powder Metallurgy

Finally, components such as automotive transmission gears and air conditioner compressor parts are manufactured through the process of POWDER METALLURGY (P/M). P/M is a process in which POWDERS OF METALS ARE MIXED ALONG WITH BINDING AGENTS, then PRESSED INTO THE SHAPE OF A MOLD, then HEATED (OR SINTERED).

Powder metallurgy allows metals that would otherwise not mix when melted to be mixed. For example, graphite does not fuse with metals by heat alone but, through P/M, graphite can be added to steel to make self-lubricating bearings.

Powder metallurgy parts are "created" much like a casting process would create a cast valve body. The main difference is that powder metallurgy uses powders that are specifically selected for the application.

The process where powders of materials are mixed along with binding agents, pressed, and heated (or sintered) into the shape of a mold is called:

casting .... **Page 8-27**
powder metallurgy .... **Page 8-28**

From page 8-28   8-31

You have now been introduced to composites, plastics, ceramics, and powder metallurgy. It should be very clear to you now that these materials do NOT fit the clear pattern of fabrication that metals do.

In fact, each ceramic, designed for a particular purpose, may well indeed contain different discontinuities. One may contain significant impurities and the next porosity.

Turn ahead to page 8-33.

From page 8-33   8-32

Oops! You missed a very important point concerning engineering materials. Let's restate our point.

It should be very clear to you now that these materials do NOT fit the clear pattern of fabrication that metals do. In fact, each ceramic, designed for a particular purpose, may well indeed contain different discontinuities. One may contain significant impurities and the next contain porosity.

Without question it is important that the nondestructive testing personnel be quite familiar with the fabrication processes of the plastic, composite, or ceramic which is to be evaluated for a particular purpose. The particular nondestructive method employed may be severely limited in its ability to assess the integrity of a product.

Turn ahead to page 8-34.

From page 8-31

Without question it is important that the nondestructive testing personnel be familiar with the fabrication processes of the plastic, composite, or ceramic which is to be evaluated for a particular purpose.

The particular nondestructive method employed may be severely limited in its ability to assess the integrity of a product. As products are shrinking in size and increasing in complexity, the nondestructive testing industry has yet many challenges ahead!

Nondestructive testing can readily accomplish the examination of any metal, plastic, composite, and/or ceramic.

True	**Page 8-32**
False	**Page 8-34**

| From page 8-33 | 8-34 |

Absolutely! Discontinuities in composites, plastics, and ceramics are NOT easily categorized.

NDT methods are under constant development and will most likely attain an acceptable level of confidence in detecting and evaluating the discontinuities associated with composites, plastics, and ceramics.

The NDT examiner should clearly know what the limitations of the applied NDT method are when dealing with these material types.

For a review of engineering materials, turn to the next page.

CHAPTER REVIEW

1. A _____ is a material that is capable of continuous and permanent shape change in any direction without breaking apart.

    A. plastic
    B. ceramic
    C. composite
    D. silicate

2. Discontinuities in composites, plastics, and ceramics are:

    A. easily categorized.
    B. not easily categorized.
    C. similar to those in the steel-making processes.
    D. always open to the surface.

3. Materials made from combinations of feldspar, silicates, clay, and bauxite are referred to as:

    A. composites.
    B. plastics.
    C. powder metallurgy components.
    D. ceramics.

4. Powders can be mixed together, heated and pressed into the shape of a mold by the process of:

   A. monomer polymerization.
   B. FRP.
   C. powder metallurgy.
   D. grain boundary direction.

5. The smallest amount of an element that can exist and still be referred to as that element is called:

   A. an atom.
   B. a molecule.
   C. a monomer.
   D. a polymer.

6. _____ are specific, repeated arrangements of atoms or molecules in a solid material.

   A. Atoms
   B. Crystals
   C. Polymers
   D. Molecules

7. Two types of plastics are:

   A. monomers and polymers.
   B. ceramics and composites.
   C. FRPs and thermosetting.
   D. thermosetting and thermoplastic.

From page 8-36

8. The engineering material, ceramics, is best known for its:

   A. ability to withstand high-temperature applications.
   B. ability to be remelted and remolded.
   C. layered structure.
   D. thermoplastic properties.

9. A material made as a layer of cloth covered by a layer of epoxy (glue) is a _____ material.

   A. plastic
   B. ceramic
   C. polymer
   D. composite

10. Assume that it is your responsibility to design a supersonic fighter aircraft wing structure. Speed, weight, and durability are essential requirements. You would most likely use a:

    A. plastic.
    B. ceramic.
    C. composite.
    D. powder metallurgy component.

Turn to the next page for answers to these review questions.

ANSWERS TO REVIEW QUESTIONS FOR CHAPTER 8

Question & Answer	Reference Page(s)
1. A	8-13
2. B	8-34
3. D	8-26
4. C	8-30
5. A	8-2
6. B	8-8
7. D	8-17
8. A	8-26
9. D	8-16
10. C	8-18

Turn to the next page.

From page 8-38   8-39

You have just completed the programmed instruction course, *Introduction to Nondestructive Testing*.

Now you may want to evaluate your knowledge of the material presented in this handbook. A set of self-test questions are included in the back of the book. The answers can be found at the end of the test.

We want to emphasize that the test is for *your own* evaluation of *your* knowledge of the subject. If you elect to take the test, be honest with yourself. Don't refer to the answers until you have finished. Then you will have a meaningful measure of your knowledge.

Since it is a self-evaluation, there is no grade and no passing score. However, if your score is 80% or higher, you probably have a good understanding of the material. If you find that you have trouble in some part of the test, it is up to you to review the material until you are satisfied that you know it.

Turn to the Self Test and begin.

APPENDIX A

INTRODUCTION

SELF-TEST

___ 1. The folding of metal in a thin plate on the surface of a forging causes:

    A. a cold shut.
    B. unhealed porosity.
    C. a lap.
    D. a seam.

___ 2. Grinding cracks in metal are always:

    A. parallel to the grinding wheel location.
    B. subsurface.
    C. open to the surface.
    D. in a checkerboard pattern.

___ 3. Where might microshrinkage occur in a casting?

    A. At the header
    B. At junctions between light and heavy sections
    C. At the gate
    D. At the casting's surface

___ 4. Severe grinding cracks appear as:

   A. wave-like patterns.
   B. widely-spaced cracks.
   C. unhealed porosity.
   D. lattice-work or checkerboard patterns.

___ 5. When a crack-like space is caused by molten metal covering solidified metal, the result is a:

   A. cold shut.
   B. lap.
   C. burst.
   D. seam.

___ 6. If a weldment has a crater crack, you might find it:

   A. at the end of the weld.
   B. at the beginning of the weld.
   C. somewhere in between the beginning and the end of the weld.
   D. at any one or all of the above places.

___ 7. Metal has the property of:

   A. failing to unite if too hot.
   B. retaining large grain structure when forged.
   C. retaining small grain structure when cast.
   D. occupying more space when molten than when solid.

8. Porosity is sometimes found in plate or sheet metal and is given the name:

    A. stringers.
    B. laminations.
    C. bursts.
    D. seams.

9. Grinding cracks occur crosswise to grinding wheel rotation.

    A. True
    B. False

10. A discontinuity can be caused within a casting as it solidifies. It would be called:

    A. a shrinkage cavity.
    B. a blow hole.
    C. an inclusion.
    D. a stringer.

11. Nonmetallic inclusions are NOT present before the billet is worked.

    A. True
    B. False

12. Which best describes what is sought by nondestructive testing?

    A. Surface irregularities
    B. Subsurface holes and cracks
    C. Any breaks in the continuity of metal structure
    D. Structure integrity

13. Seams are subsurface discontinuities.

    A. True
    B. False

14. Which of the items listed below are discontinuities found in ingots?

    A. Laminations
    B. Bursts
    C. Crater cracks
    D. Nonmetallic inclusions

15. Stringers are a type of discontinuity sometimes found in:

    A. sheet metal.
    B. bar stock.
    C. forgings.
    D. castings.

16. Oxide in weldments can cause discontinuities that are similar to nonmetallic inclusions, but these oxide-caused discontinuities are called:

   A. crater cracks.
   B. seams.
   C. slag inclusions.
   D. bursts.

17. Heat treating might cause discontinuities that would probably start at:

   A. heavy and light junctions and sharp areas.
   B. smooth surfaces.
   C. concave surfaces.
   D. the outside of the part.

18. Blow holes occur beneath the surface of castings.

   A. True
   B. False

19. Forging laps can occur in which of the places described below?

   A. The top of the die
   B. Were the dies come together
   C. The bottom of the die
   D. Within the die

20. If a weld is made and the parent metal is restrained, a crack might occur due to stress. It would usually occur:

   A. in the direction of the weld.
   B. in the root of the weld only.
   C. only in the parent metal.
   D. transverse to the direction of the weld.

21. Which of the following is present in ingots?

   A. Porosity
   B. Stringers
   C. Seams
   D. Laminations

22. In its original shape, porosity would appear:

   A. elongated.
   B. irregular.
   C. round or nearly round.
   D. square.

23. Nondestructive testing is a method of locating which of the following?

   A. Discounts
   B. Dislocations
   C. Discontinuities
   D. Disturbances

24. Like ingots, welds can contain porosity.

    A. True
    B. False

25. Nonmetallic materials trapped in solid metal are described as:

    A. porosity.
    B. inclusions.
    C. seams.
    D. laminations.

26. A hot tear in a casting is a crack. It could occur:

    A. within the heavy part of the casting.
    B. in the light section of the casting.
    C. at junctions of light and heavy sections.
    D. as a subsurface discontinuity.

27. Many small, subsurface holes can result when a pouring gate in a casting blocks off prematurely. This condition is called:

    A. hot tears.
    B. crater cracks.
    C. microshrinkage.
    D. seams.

28. A forging lap might occur:

    A. were grain direction makes an abrupt change.
    B. in the center of the forging.
    C. at a junction of light and heavy sections.
    D. as a subsurface discontinuity.

29. A lamination might be found in:

    A. bar stock.
    B. "pipe."
    C. forgings.
    D. sheet or plate metal.

30. In the beginning, all steel is perfect.

    A. True
    B. False

31. What is a nonmetallic inclusion called in bar stock?

    A. Lamination
    B. Burst
    C. Lap
    D. Stringer

____ 32. Nonmetallic inclusions are:

   A. round in shape.
   B. rectangular in shape.
   C. irregular in shape.
   D. tall.

____ 33. Grain direction in steel bar stock is always:

   A. through the length of the bar.
   B. through the width of the bar.
   C. spread out in all directions.
   D. A bar of steel has no grain.

____ 34. In forgings, grain:

   A. follows the shape of the dies.
   B. becomes coarser.
   C. retains its original shape.
   D. does not exist.

____ 35. In which of the following might you expect to find a seam?

   A. Forgings
   B. Castings
   C. Bar stock
   D. Sheet or plate metal

36. A cold shut has a:

    A. smooth, curved appearance.
    B. ragged shape.
    C. straight, thin shape.
    D. straight, thick shape.

37. Premature blocking of the entrance from a light to a heavy section during the pouring of a casting might cause:

    A. a lap.
    B. microshrinkage.
    C. stringers.
    D. bursts.

38. Nondestructive testing is used for locating only those discontinuities which are open to the surface.

    A. True
    B. False

39. A material that is composed of layers of other materials is referred to as:

    A. a plastic.
    B. a ceramic.
    C. layered materials.
    D. a composite.

40. Given all materials discussed, nondestructive testing _____ of discontinuities.

    A. can readily assess all types
    B. can only find surface types
    C. is limited in detection and evaluation
    D. will only find subsurface types

41. _____ are materials that can be readily reheated and remolded.

    A. Plastics
    B. Ceramics
    C. Monomers
    D. Molecules

42. _____ are defined as the smallest unit of a compound made of two or more atoms grouped together by chemical bonds.

    A. Atoms
    B. Molecules
    C. Plastics
    D. Thermosets

43. The two types of plastics are:

    A. thermosetting and monomer.
    B. thermoplastic and monomer.
    C. thermosetting and thermonuclear.
    D. thermoplastic and thermosetting.

44. The two basic categories for pipes and tubes are _____ and _____.

   A. welded, cast
   B. welded, seamless
   C. seamless, forged
   D. cast, forged

45. A discontinuity caused by lack of fusion in the weld on a welded tube is a:

   A. stringer.
   B. weld spot.
   C. seam.
   D. de-fusion.

46. Extruded parts may contain any of the discontinuities that were in the _____ from which they were formed.

   A. weld
   B. mandrel
   C. seams
   D. stock

ANSWERS FOR SELF-TEST

Question & Answer	Reference Page(s)	Question & Answer	Reference Page(s)
1. C	3-7	24. A	7-23
2. C	6-1	25. B	1-28
3. C	4-31	26. C	4-6
4. D	6-1	27. C	4-30
5. A	4-4	28. A	3-16
6. D	7-3	29. D	2-6
7. D	4-24	30. B	1-26
8. B	2-6	31. D	2-15
9. A	6-1	32. C	1-35
10. A	4-24	33. A	2-2
11. B	1-35	34. A	3-3
12. C	1-15	35. C	2-17
13. B	2-18	36. A	4-6
14. D	1-42	37. B	4-30
15. B	2-15	38. B	1-20
16. C	7-25	39. D	8-17
17. A	6-4	40. C	8-30
18. B	4-38	41. A	8-12
19. B	3-7	42. B	8-2
20. D	7-15	43. D	8-14
21. A	1-42	44. B	5-1
22. C	1-36	45. C	5-3
23. C	1-23	46. D	5-22

APPENDIX B

GLOSSARY

Atom   The smallest amount of an element that can exist with the properties of that element.

Base Metal   The metal that is to be welded.

Billet   The material shape after the bloom has been processed from a cropped ingot.

Binding Forces   The forces that hold atoms and molecules in a particular arrangement.

Bloom   The name of the ingot after cropping. Physically larger than a billet.

Blow Holes   Holes blown into a casting's surface by gases expanding in the mold itself.

Burr   A sharp edge that remains after forming or shaping metal.

Castings   Made by the pouring of liquid (molten) metal into a mold.

Ceramics   Compounds of oxygen, nitrogen, silicates, and carbon that have been formed through high-temperature processing.

Cold Shuts   Formed when molten metal splashes onto and adheres to the surface of previously solidified metal in castings.

Composites   Materials made up of distinct parts, usually one material layered onto another, with each material keeping its own identity.

Crater Cracks   Cracks occurring in the weld crater as a result of improper use of heat.

Crater   Formed by improper welding technique when a weld is started or stopped.

Cropped   The removal of the hot top of an ingot.

Crystal   A solidified material in which the atoms or molecules are arranged in a definite repeating pattern.

Defects   Undesirable aspects of metal processing; unacceptable discontinuities.

Destructive Tests   A means of testing specific properties of an article by changing the physical qualities in such a way as to destroy the usefulness of the article.

Die   A tool or device used for the shaping or forming of metals and other materials.

Discontinuity   A break or interruption in the normal physical structure of an article.

Eddy Current Testing   A nondestructive testing method that uses an electrical current in a coil to induce eddy currents into a conductive specimen. Discontinuities alter the path of the induced currents.

Explosive Forming   Metal forming process that uses an explosive charge to shape the article into a die or mold.

Extrusion   The forming of parts by forcing metal through a die.

Fatigue   Weakening of metal parts through repeated cyclic use.

Feeder Head   A reservoir that allows additional metal to flow into a casting to replace the solidifying and shrinking metal.

Ferrous   Of or pertaining to iron.

Forging   The working of metal into a desired shape by hammering or pressing the metal while it is very hot and in a malleable condition.

Forging Bursts   A rupture caused by forging at improper temperatures. Can be internal or external.

Forging Laps   A discontinuity caused by the folding of a thin piece of metal on the surface of the forging; forging laps are always open to the surface.

Gate   Entrance to the mold through which the molten metal is poured (castings).

Gouging   Cavity caused by friction between the mandrel and the inside surface of the pipe or tube.

Grain Structure   The structure of the grains that give a material its mechanical properties (i.e., tensile strength, hardness, toughness, etc.).

Grinding Cracks   The result of the heating, expanding and cooling of metal in the grinding process. These cracks occur at a right angle (crosswise) to grinding wheel rotation.

Heat-Affected Zone A narrow part of the base metal on each side of the weld that has been affected by the heat generated while welding.

Heat Treatment The process of hardening or softening certain metals by controlled heating and cooling. Desired mechanical properties, such as machinability, can be introduced through this process.

Hot Top The top of the ingot, often discarded, used to absorb most of the ingot shrinkage.

Hot Tears Separations (rips) in metal from unequal cooling (castings).

Inclusion Unwanted impurities within the metal weld.

Indication Evidence as a result of a nondestructive test.

Ingots The solidified metal that has been removed from the initial steel-making pouring mold. Ingots can be many sizes.

Intermittent Occurring at irregular intervals.

Lack of Fusion (LOF) Failure of the molten metal to fuse with the adjacent (often base) metal.

Lack of Penetration (LOP) Failure of the molten weld metal to fuse with the base metal at the bottom or base (root) of the weld joint.

Lamination Inclusions in plate material that have been flattened out in all directions, but mainly in the direction of roll.

Line of Fusion The area between the base metal and the newly-formed weld.

Liquid Penetrant Testing A nondestructive testing method that uses a penetrating liquid to seep into surface discontinuities ultimately providing a visible indication.

Longitudinally Oriented with the long axis of a pipe, plate, bar, or weld.

Magnetic Particle Testing A nondestructive testing method that uses electrical current to create a magnetic field in a specimen while applied magnetic particles indicate where the field is broken by discontinuities on or near the surface.

Microshrinkage Many small (microscopic) subsurface cracks or cavities.

Molecule The smallest unit of a compound that retains the properties of that compound, made of two or more atoms grouped together by chemical bonds.

Monomers Means one part, and refers to individual molecules used to define a compound (e.g., plastics).

Nonmetallic Inclusions Pockets of encapsulated impurities (slag).

Nondestructive Testing Methods of testing articles (parts or materials) for discontinuities, including cracks or flaws, without damaging the usefulness of the article.

Piercing Mandrel A plunger that penetrates through the length of a bar, forming a rough pipe without any seam.

Pig Iron Iron containing impurities that has been poured into molds (resembling suckling pigs).

Pipe A discontinuity caused by molten metal shrinking when it cools and solidifies; a cylindrical tube.

Plastics Man-made nonmetallic substances that can be shaped or molded under heat and/or pressure.

Polymerization The process of linking monomers together.

Porosity Entrapped gas in solidifying metal (casting or welding).

Powder Metallurgy (P/M) A process in which powders of metal are mixed with binding agents then pressed into a mold and heated (sintered).

Quality Assurance Planned and systematic actions taken to determine if an article meets the required characteristics and will perform as expected.

Radiographic Testing A nondestructive testing method that uses electromagnetic rays (X-rays and gamma rays) to penetrate material(s), recording discontinuities in the material(s) on film. Detects both surface and subsurface discontinuities.

Seamless Having no seams (i.e., seamless pipes and tubes).

Shrinkage Cracks See **Hot Tears**.

Shrinkage Cavities Subsurface holes caused by the contraction of the cooling and solidifying of metal.

Skelp A flat piece of steel from which welded tube is made.

Slabs A processed bloom.

Slag Inclusion Inclusions made of slag that can occur in the weld bead.

Slag The impurities remaining from the smelting or welding processes.

Slugs Pieces of metal fused into tube or pipe (usually from the piercing mandrel).

Smelting A melting process used for separating pure metal from impure substances.

Stress Pressure exerted on a body that tends to deform its shape.

Stringer Inclusions in blooms or billets that have been stretched out, mainly in the direction of roll.

Subsurface Discontinuities Discontinuities within a material not exposed to the surface.

Surface Discontinuities Discontinuities exposed to the surface.

Thermoplastic Plastics that can be remelted and remolded as a result of end-on-end monomer chains.

Thermosetting Plastics that cannot be remelted and remolded due to cross-linking of monomer chains.

Transverse Oriented across the main axis of a pipe, plate, bar, or weld.

Ultrasonic Testing A nondestructive testing method that uses ultrasound to penetrate and evaluate materials. Detects both surface and subsurface discontinuities.

Undercut Melting away of the base metal adjacent to the weld caused by improper welding technique.